EATING IN US NATIONAL PARKS

This book presents a fascinating exploration of eating experiences within US national parks, explaining how, on what, and why people eat in national parks and how this has changed over the last century.

National parks are enjoying unprecedented popularity, and they are especially popular sites for the expression of cosmopolitanism, an ideological outlook descended from the Romantics on whose vision the parks were originally founded. The book explores the constructed food-scape within US national parks, situating the romantic consumption ethos within the context of sociological work on distinction, culinary tourism, and culinary capital. It analyzes and problematizes elements of cosmopolitan taste and desire, examining food tourism in wilderness spaces that satisfies cosmopolitan hunger for authenticity and a certain type of self-making. Weaving together strands of research that have not been previously integrated, the book gleans meaning from concessions menus and park restaurant web pages and employs audience analysis to take stock of park restaurant visitors' contributions to restaurant review websites, as well as to understand how they represent their park eating experiences on social media. The book examines how satisfying cosmopolitan tastes in the parks creates profit for corporate concessioners, but also may produce bioregionalist successes and a recentering of Indigenous foodways. It concludes by exploring inroads to a better food experience in the parks, involving food products and processes that are regionally/locally specific, where tourists witness and participate in food production and enjoy commensality, but that are also non-extractive and show care for the environment and the people who inhabit it.

This book will be of great interest to students and scholars of food studies, tourism and hospitality, sociology of culture, parks and recreation, American studies, and environmental studies. The book will also be of interest to parks and recreation decision makers, sustainable tourism leaders, and hospitality managers.

Kathleen LeBesco is Professor of Communication and Media Arts and Associate Vice President for Strategic Initiatives at Marymount Manhattan College, USA. She is coauthor/coeditor of multiple books, including *The Bloomsbury Handbook of Food and Popular Culture* (2017), *Culinary Capital* (2012), and *Edible Ideologies* (2008). She is a former snack bar technician, line cook, and restaurant reviewer for *Time Out New York's Eating and Drinking Guide*.

Routledge Food Studies

Food Education and Gastronomic Tradition in Japan and France
Ethical and Sociological Theories
Haruka Ueda

Finding Meaning in Wine
A US Blend
Michael Sinowitz

Food Policy and Practice in Early Childhood Education and Care
Children, Practitioners, and Parents in an English Nursery
Francesca Vaghi

Eating in US National Parks
Cosmopolitan Taste and Food Tourism
Kathleen LeBesco

For more information about this series, please visit: www.routledge.com/Routledge-Food-Studies/book-series/RFOODS

EATING IN US NATIONAL PARKS

Cosmopolitan Taste and Food Tourism

Kathleen LeBesco

LONDON AND NEW YORK

from Routledge

Designed cover image: © Getty

First published 2024
by Routledge
4 Park Square, Milton Park, Abingdon, Oxon OX14 4RN

and by Routledge
605 Third Avenue, New York, NY 10158

Routledge is an imprint of the Taylor & Francis Group, an informa business

© 2024 Kathleen LeBesco

British Library Cataloguing-in-Publication Data
A catalogue record for this book is available from the British Library

ISBN: 978-1-032-59632-7 (hbk)
ISBN: 978-1-032-59631-0 (pbk)
ISBN: 978-1-003-45551-6 (ebk)

DOI: 10.4324/9781003455516

Typeset in Times New Roman
by codeMantra

For Evan Franz, unrelenting
In Memoriam

CONTENTS

ACKNOWLEDGMENTS

While writing this book, I benefited from the support and good company of many colleagues at Marymount Manhattan College including Sue Behrens, Peter Naccarato, Brian Rocco, Desiree Sholes, Jill Stevenson, Laura Tropp, Tunisia Wragg, Emmalyn Yamrick, and Diana Zambrotta Sheetz. I am especially grateful to students Martha Madrid and Sheridan Poschelle in Prof. Diana Epelbaum's Race and Place in the Natural Histories of the Americas class, whose presentations spurred me to think seriously about Indigeneity and eating in national park spaces. Librarians Mary Brown and Jason Herman provided invaluable assistance with endless Interlibrary Loan requests. I am indebted to David Podell and Judson Shaver for granting me, more than a dozen years ago, the research sabbatical that would eventually allow me to take a belated break from college administration to complete the writing of this book and to Kerry Walk for encouraging me to actually do it.

This book was developed out of many conference presentations across a range of disciplines. I was fortunate to present parts of this project at the 2019 and 2021 conferences of the Association for the Study of Food and Society, the 2020 and 2022 International Conferences on Food Studies, the 2021 Second Biennial Conference on Food and Communication, the 2021 Food History Seminar of the Institute of Historical Research, and the 2022 Critical Tourism Studies IX Conference. I appreciate the labor of those who organized these meetings, as well as the comments and questions of those scholars who listened attentively and who helped me to strengthen my work.

I benefited from close readings of early chapters by members of a paper incubator group organized by the Association for the Study of Food and Society: Stephanie Borkowsky, Shayne Figueroa, Malia Guyer-Stevens, Paolina Liu, and Katherine MacGruder offered gracious suggestions for improvement, and many laughs, throughout 2021 and 2022. Other buddies in and around the food studies universe enthusiastically allowed me to bend their ears about this project. Thank you, Kathleen Collins, Christine Gregory, and Aiko Tanaka.

Countless friends have accompanied me on park trips in the years since I started this project and endured various iterations of the elevator pitch for this book. Your feedback and questions helped me to sharpen my argument. Thanks to Andria Alefhi Lamberton (Big Bend); Susan Ericsson and Scott Perry (US Virgin Islands, Indiana Dunes, Saguaro, Yellowstone, Crater Lake, Redwood, Mt. Rainier, Olympic, and North Cascades); Meg Honsinger (Congaree, Rocky Mountain, Black Canyon of the Gunnison, Mesa Verde, and Great Sand Dunes); and Nancy Inouye, Kai Inouye-Merritt, and Taro Inouye-Merritt (Yellowstone, Redwood, Death Valley, and Joshua Tree) for robust hiking company, for your unending patience with my need to interrogate snack bar menus, and for your ability to stomach endless cafeteria grilled chicken sandwiches with good cheer.

At Routledge Press, I appreciate Hannah Ferguson for her editorial acumen and Katie Stokes for helping me manage the nuts and bolts of this project. I also appreciate the comments of the anonymous reviewers secured by the Press; their feedback helped me to focus my arguments and tighten my writing.

Last but not least, I would like to thank my family. Molly and John Shields, you have schlepped along on more park trips than you probably care to remember at this point. In our little house, you gave me space to write during the height of the COVID pandemic, when having a meaningful long-term project to work on helped keep me sane. And every day, you create a warm home filled with laughter and light that allows me to thrive. Thank you.

INTRODUCTION

In late March 2021, after a full year in which a masked walk around my own block was the height of my journey-making, I got vaccinated against COVID-19. Two weeks and one day later, I hopped a plane, bound for the national parks of New Mexico and West Texas, where I welcomed the opportunity for a change of scenery, but also an outdoor vacation where I wouldn't have to be indoors with the rest of my potentially germy fellow humans. Something of a national park geek, I had been following with interest stories about dramatic increases in park usage when the parks were open—many were fully closed for months during COVID, according to National Park Service (NPS) annual visitation statistics (National Park Service 2021).

At New Mexico's Carlsbad Caverns, as I stood by myself, awaiting the NPS ranger-led orientation that would then allow the day's first group of hikers to hit the steep trail down into the caves, I couldn't help but overhear the conversation of the group of attractive, athletic thirty-somethings in front of me, which was about lunch. They were debating how long to spend hiking in the caverns, when to eat lunch, and what to have. One woman pointed to the cafe behind us in the Visitor Center, which serves standard park fare—a lot of shelf-stable comfort food—and suggested they grab a bite there midday after they returned to the surface. A guy in head-to-toe Patagonia gear muttered, "Ugh, but the line is going to be so long—these parks hardly have anyone working at them because of COVID." Someone else in the group chimed in, "Yeah, and the food is SO. NOT. INTERESTING. It's, like, a sad quesadilla and some iceberg lettuce. For like $15." Patagonia guy then said that he had looked at Culture Trip, a website for restaurant recommendations, and thought maybe they should head into Carlsbad proper for some barbecue, advising his friends: "It's supposed to be authentic, it's a hole-in-the-wall place." The attractive thirty-somethings nodded along and agreed to grab a snack down in the

DOI: 10.4324/9781003455516-1

cave—presumably at the snack bar-style Underground Lunchroom—to tide them over until they could get to the barbecue joint.

I listened with great interest to this conversation not only because I'm nosy but because their exchange sounded familiar notes of concern over the role of the concessions system in the twin goals—conservation and enjoyment by the public—of the NPS and reflected something interesting about the identity of the speakers and their hopes for something better, more real, and interesting, to eat in their wilderness travels.

A Personal History

I grew up with neither awareness of nor interest in national parks. My blue-collar family was not outdoorsy, and we did not take vacations; instead, we were screen people. Inspired by a viewing of the film *Grand Canyon* and some cactus-filled brochures from a possible graduate school destination out west, I made my first pilgrimage to a national park in my very early 20s. But the South Rim of the Grand Canyon was fogged in, and after flying cross-country and then driving six hours through more weather systems than I thought could converge in a single day, I could not even see my own hand in front of my face. Foiled, I meandered around the visitor center, browsed in a gift shop, got back in the car, and drove back to my motel in Tusayan, with its view of outsized, roadside Flintstones character billboards. Stone age cartoon figures replaced the majestic rocks I had been pining to see—"yabba dabba *don't*," they taunted. The next morning, the weather had not lifted, and so I slogged back to Tucson with my tail between my legs. I wouldn't get back to the Grand Canyon for almost 25 years.

It was that first encounter with a national park, during which I literally could not see the land, that sparked my desire to look more deeply. As a full-fledged grown-up, and now a parent, I have come to regard national park travel as the holy grail of family vacations—Beautiful landscapes! Fresh air! No screens! I have also meditated on what national park travels make plain and what they obscure. During this time, I have been working in food studies, and I have noticed surprising juxtapositions in my travel experiences. Accessing some of the higher-end dining experiences in the parks bears more than a passing resemblance to getting a table at a hot restaurant in NYC, the city where I've worked (and grazed) for 25 years: a long-term reservations strategy is required, which requires a certain amount of cultural and culinary capital. Most strikingly, though, the dining scene in today's parks is dominated by industrial concessions that seem a far cry from the romantic consumption imagined by early park advocates.

Overview

If national parks were meant to be places for transformative experiences of the sublime, why isn't the food better? This question is not just about how the food tastes but about its impact on culture and environment. How did we get to this moment where park eating is dominated by industrial concessions, and what is this foodscape doing to our people and our land? How are people using global flows of media to assert tastes and make demands that we do better with food on our public lands? And how are foodways increasingly taken up in the interpretive mission of the national parks?

National parks are enjoying unprecedented popularity, and they have a long heritage as spaces for soul-enhancing outdoor recreation and peaceful contemplation. They are especially popular sites for the expression of cosmopolitanism, a set of attitudes and practices that descend from the Romantics on whose vision the parks were originally founded. Cosmopolitanism prioritizes experiences over things and valorizes consuming romantically. US national parks, which offer visitors opportunities to experience nature's sublimity and to enjoy wholesome and meaningful travel, fit the bill. However, travelers to the parks today are confronted with a food system that seems in direct tension with the mission of the parks and that complicates the expression of cosmopolitan identity. That makes the eating experiences in the parks fascinating sites for analyzing struggles over status and identity and considering what is at stake in the offing.

The people who visit national parks shape the food system there, and that same food system has shaped local cultures and landscapes alike. This manuscript explores the constructed foodscape within US national parks, situating the romantic consumption ethos within the context of sociological work on distinction, culinary tourism, and culinary capital, pointing the way to an explanation of how, on what, and why visitors dine in national parks and how this has changed over the last century. I analyze and problematize elements of cosmopolitan taste and desire, examining food tourism in "wilderness" spaces that satisfies cosmopolitan hunger for authenticity and a certain type of self-making. Although there are dangers in catering to cosmopolitan tastes, these tastes also create sustainable possibilities for the lands and the people who inhabit them.

Methodologically, I build on these intertwined strands with an original investigation using a mixed methods approach. I use textual analysis to glean meaning from concessions menus and park restaurant web pages. I employ audience analysis to take stock of park restaurant visitors' contributions to restaurant review websites, as well as to understand how they represent their park eating experiences on social media. I also employ political economy analysis when I examine how the satisfaction of cosmopolitan tastes in the parks creates profit for corporate concessioners but also furthers progressive causes of bioregionalism and Indigenous food sovereignty.

Outline of Chapters

This book provides an analysis of how cosmopolitan self-making shapes and is shaped by national parks. The cosmopolitan tastes of food-minded travelers reinforce taste hierarchies, allowing some to benefit from the mantle of "good taste" that their cultural capital confers. However, these same tastes may support environmental and cultural repair. Chapter 1, "Tourism and Taste: From Romanticism to Cosmopolitanism," applies concepts of cosmopolitanism, taste, and food tourism to the landscapes of the national parks. The chapter connects the historical influence of the Romantics on the national parks to the contemporary desires of cosmopolitan food tourists, defines different modes of cosmopolitanism, situates the parks as challenging spaces for virtuous consumption due to the presence of corporate concessioners, and previews the promise of pragmatic cosmopolitanism for the land and its people.

Chapter 2, "Industrial Food in the Wilderness: Dining and Democracy," provides an analysis of how tensions between dining and democracy play out in the history of park concessions. The chapter traces historical anxieties about taste and tackiness in Niagara Falls and Coney Island that informed the development of a more restrained and "civilized" version of concessions in the national parks. I look closely at the rise of luxury concessions within what would become Yosemite National Park and demonstrate how these concessions both responded to and shaped cosmopolitan tastes. I also examine the role of early NPS leadership in shaping the public imagination about park visitors as the kind of educated, curious citizens that would come to align with cosmopolitan identity. The chapter concludes by exploring changing federal concessions regulations that have intermittently imperiled and enabled the full expression of the tasteful tourist's identity.

In Chapter 3, "Indigeneity and Eating in US National Parks," I explore a cultural history of Indigenous foodways in what are today's parklands. Here, I detail struggles for American Indian food sovereignty on productive foodscapes that were appropriated by settler colonists, ultimately to be reimagined as uninhabited wilderness and made into national parks. I analyze the invention of the idea that parks are spaces of pristine wilderness, exploring its usefulness to notions of Romantic self-hood and its harm to Indigenous people and their foodways. Specifically, I present instances in which Native Americans were prevented from hunting, fishing, or harvesting in order to conserve parklands alongside evidence of the aversion park advocates had to Indigenous people's food and general decorum. I document the active role played by the US federal government in transitioning many Indigenous people to reservation-based agriculture, and ultimately to hunger and dependency, by dispossessing them from their lands and their traditional foodways. I situate all this as an important enabling condition of white cosmopolitan self-making, while exploring emerging efforts by Native Americans to re-center their foodways relative to national parks.

Chapter 4, "Swallowing Tensions: Exploring the Contemporary Foodscape," looks at the contemporary food concessions system in the parks and presents a case study of the representational rhetoric of one large corporate concessioner across a range for restaurant types. I provide an overview of the concessions options available in today's parks, and I detail the relationship between NPS requirements and concessioner operations, particularly with regard to the fulfillment of the parks' interpretive mission. The chapter then delves deeply into an analysis of the public image and communications of Xanterra, a major park concessioner, across a range of its restaurants in Yellowstone National Park. Through a close read of restaurant menus and concessioner web pages, I show how tensions between conservation and enjoyment that have been inherent in the parks since they were founded play themselves out as travelers are encouraged to imagine their individual acts of romantic consumption as supporting a better and more sustainable foodscape.

In Chapter 5, "Representing Upscale Restaurants," I analyze the narratives provided by several large park concessioners in selling their most expensive meals, demonstrating how upscale park restaurants, redolent with symbolic value, are sites for the negotiation of status among eaters. I explore how digital media focused on upscale park restaurants are both reflecting and shaping discourses of health, sustainability, the local, landscape, and cosmopolitanism today. Focusing on examples from Acadia National Park's Jordan Pond House, Mesa Verde National Park's Metate Room, and Grand Teton National Park's Jenny Lake Lodge Dining Room, I show how health discourses are underplayed relative to indicators of sustainability and local foodshed, which are attended to more vigorously by concessioners and their customers. I also examine the integration of visual vocabularies of Indigeneity into these high-end eating experiences and show how an emphasis on landscapes dominates representations of the most valued dining experiences in the parks.

Finally, Chapter 6, "Reimagining Food in National Parks: Future Ecologies of Bioregionalism and Indigenous Food Sovereignty," examines future ecologies of bioregionalism and Indigenous food sovereignty. I draw examples from across the US parks system to make sense of the gestures toward place-based gastronomy by corporate concessioners, the presence of park-adjacent culinary trails that emphasize local heritage cuisine for purposes of regional economic development, and the existence of national parks that incorporate heritage orchards or working farms as an element of land conservation. I also analyze, evaluate, and propose some creative ways that food is being reimagined in the parks by Indigenous people and their supporters in the NPS. This conclusion explores inroads to a better food experience in the parks, involving food products and processes that are regionally/locally specific, where tourists witness and participate in food production and enjoy commensality, but that are also nonextractive and show care for the environment and the people who inhabit it.

Reference

National Park Service. 2021. "Annual Visitation Highlights." Last updated April 8, 2021. https://www.nps.gov/subjects/socialscience/annual-visitation-highlights.htm.

1

TOURISM AND TASTE

From Romanticism to Cosmopolitanism

Romantic sensibilities have significantly shaped the past of the national parks, and cosmopolitan tastes stand to shape their future. Romanticism, a movement that emphasized individual self-expression, emotion, imagination, nature, and the personal experience of the sublime (Swiggett 1903), gained traction in nineteenth-century Europe as a critique of the Enlightenment rationality that had ushered in the Industrial Revolution. Cosmopolitanism is a contemporary cultural discourse that values global perspectives, openness, and engagement with a range of cultural practices and groups (Skey 2012); it is often understood as a critique of nationalism, localism, and other parochial perspectives. These ideological frameworks animate experiences of travel and tourism in general and of eating in national parks in particular.

Far from the Madding Crowd: Romantic Contemplation, Nature, Gastronomy, and Taste

In 1865, Frederick Law Olmsted, best known as the father of US landscape design and landscape designer for New York's Central Park, was an early key advocate for national parks. Influenced by Romanticism, Olmsted believed that the aesthetic experience of beautiful landscapes could free the imagination (Fisher 1986). Olmsted authored a report—a blueprint for the fitting together of travel, contemplation, class, and aesthetics—about his designs for the land that would later (in 1890) become Yosemite National Park.

In the report, Olmsted waxes rhapsodic in a manner bordering on purple prose about the features of the land there. Describing a stand of stately giant sequoias, Olmsted writes:

DOI: 10.4324/9781003455516-2

there are hundreds of such beauty and stateliness that, to one who moves among them in the reverent mood to which they so strongly incite the mind, it will not seem strange that intelligent travelers have declared that they would rather have passed by Niagara itself than have missed visiting this grove.

(Olmsted 1865, np)

Olmsted swipes at the so-called less intelligent travelers who would flock to Niagara Falls, overrun by industry and commerce, framing visitors to Yosemite as distinct in their reverence.

Olmsted argued for the preservation of the Yosemite land and its consecration as a national park largely on the grounds of the civilizing effect that appreciating its beauty could have.

The power of scenery to affect men is, in a large way, proportionate to the degree of their civilization and to the degree in which their taste has been cultivated. Among a thousand savages there will be a much smaller number who will show the least sign of being so affected than among a thousand persons taken from a civilized community. This is only one of the many channels in which a similar distinction between civilized and savage men is to be generally observed. The whole body of the susceptibilities of civilized men and with their susceptibilities their powers, are on the whole enlarged. But as with the bodily powers, if one group of muscles is developed by exercise exclusively, and all others neglected, the result is general feebleness, so it is with the mental faculties. And men who exercise those faculties or susceptibilities of the mind which are called in play by beautiful scenery so little that they seem to be inert with them, are either in a diseased condition from excessive devotion of be mind to a limited range of interests, or their whole minds are in a savage state; that is, a state of low development. The latter class need to be drawn out generally; the former need relief from their habitual matters of interest and to be drawn out in those parts of their mental nature which have been habitually left idle and inert.

(Olmsted 1865, np)

Olmsted says in the report that rich people can buy land that inspires them and notes that they have plenty of opportunities to see this type of scenery for themselves, but he thinks the effects are so favorable for everyone that the government has a political duty to provide "great public grounds for the free enjoyment of the people under certain circumstances" (1865, np). He argues that "humble toilers" involved in "almost constant labor" have historically been provided by the governing classes with artificial pleasures for recreation, and that this has given them the appearance of "dullness and weakness and disease." In contrast, he argues for the preservation of natural scenery so that even the humble toilers can stand to be transformed—civilized—by their appreciation of its beauty: "It is an important fact that as civilization advances, the interest of men in natural scenes of sublimity and beauty increases" (Olmsted 1865, np). Scenery could, in Olmsted's mind, elevate popular taste (Sax 1980, 21), rather than just serving it. In this way, the national parks were imagined as tools for civilizing the uncivilized. Nature travel could be challenging in Olmsted's day, but its transformative power was unparalleled.

Romantics valorized highly aestheticized personal experiences and, in doing so, created the context into which gastronomy could emerge. Commonly understood as "the art of the table," the concern of gastronomy was defined in 1825 as "the preservation of man by means of the best possible food"; it was concerned with taste and "the action of food [...] on the moral of man, on his imagination" (Brillat Savarin 1825, Meditation III). Romanticism coincided with the invention of the restaurant and a certain type of food writing and emblematized the sophistication and social positioning associated with modern gastronomy (Gigante 2007, section 1). Gastronomes around the turn of the nineteenth century made "fine art of food" and "crusaded for the value of the aesthetic in an age of increasing consumerism" (Gigante 2007, section 1). The culture of gastronomy awakened by Romanticism took food seriously as a canvas for the expression of aesthetic judgments and a site for flexing one's intellect (Gigante 2007, section 9).

Travel has helped people to develop the ability to discern between environments that are aesthetically pleasing and unspoiled and those that are not (Urry 1995), and food is a central area of heightened discernment. Travel thus emerges as an important site for the cultivation of taste and the valorization of some tastes and experiences over others, and food in the context of travel is perhaps its apotheosis. Sociologist Pierre Bourdieu wrote,

> The opposition between the immediate and the deferred, the easy and the difficult, substance (or function) and form, which is exposed in particularly striking fashion in bourgeois ways of eating, is the basis of all aestheticization of practice and every aesthetic.
>
> *(Bourdieu 1984, 196)*

With their remoteness and visual splendor, the national parks are obvious sites of aestheticization, and the foodways there are animated by tensions between easy and difficult, form and function, and the immediate and the deferred.

The movement for national parks in the US, which took flight in the second half of the nineteenth century, just on the heels of the dawn of transcontinental railroad travel, exemplified Romantic sensibilities that intertwined religious truth and nature. The Romantics believed that people could, through introspection and observation of nature, experience the sublime and discover the divine (Campbell 1987, 182–3). Wilderness landscapes were culturally desirable because they provided spaces for this kind of experience, and because the challenging nature of accessing them reflected well on the traveler. Visiting the wilderness was, for people of a certain social standing, a desirable form of leisure activity and an expression of good taste (Urry 1995, 213). Following Olmsted's lead, the establishment of national parks and the development of infrastructure that reduced the difficulty of getting to them helped to cultivate the interests of a wider swath of humanity beyond the moneyed elite. What people ate while on these Romantic quests to parklands was the consequence of federal regulation, political and cultural disenfranchisement, and a changing national and regional food system, and had the power then (as it does still today) to convey something about who they are and what they value.

Transporting Taste

Before the twentieth century, it was the upper classes who could afford to travel to the few existing national parks. To reach Yosemite Valley in the 1860s, "it was necessary to take a boat from San Francisco to Stockton, followed by a sixteen-hour stagecoach ride to Coulterville, and finally a fifty-seven-mile, thirty-seven-hour trek by horse and pack mule into the valley" (Sax 1980, 5). In the 1870s, the best route to reach Yellowstone required

> a steamboat up the Missouri River 400 miles to the Yellowstone River, up that another 360 miles to the mouth of the Bighorn, then another 60 miles up the Bighorn to Clarks' Fork. At this point a coach would take travelers the last 72 miles to the park's border.
>
> *(Zaslowsky and Watkins 1994, 18)*

To cover the thousand or so miles from Bismarck, North Dakota, to Yellowstone, it took at least three weeks and cost $100 (Zaslowsky and Watkins 1994, 18), a princely sum at the time.

The train and the transcontinental railroad, developed in the US around the same time that the idea of creating national parks took hold, democratized and changed the character of travel. Passing through landscapes that they observed without otherwise engaging, travelers became tourists when riding the train. "To be a tourist meant to be divorced from the realities of any visited place, to re-create its essence in the context of the cultural baggage a traveler brought along" (Rothman 1998, 39). The capacity for reflective judgment that has come to serve today's cosmopolitan travelers well was seeded by the Romantics, nurtured by the Victorians, and came of age on the trains. Writing of the Victorians, Feifer noted that "the more developed one's sensitivity, the more one would want to linger over the landscape" (1985, 167). Although recreating in nature was long associated with aristocratic privilege (Meeker 1973, 4), transportation networks developed in the mid-nineteenth century would extend the privilege more widely. Similarly, the development of restaurant dining in the nineteenth century "made eating a form of entertainment and an object of conspicuous consumption as well as sustenance" (Lobel 2014, 6).

Trains allowed mass transit to the parks at least in theory, starting in 1901 when the Grand Canyon Railway rolled into the South Rim. No longer was the $15, eight-hour stagecoach ride from Flagstaff the only game in town—now, visitors could pay $3.95 for a much shorter train ride (History of the Train 2022). However, their ticket prices were still out of reach for most Americans. With the trains in play, simply going anywhere at all was no longer a marker of elite status; rather, where one went and what one did there provided a more important badge of distinction. One's destination became socially significant, and working-class resorts were branded and disregarded as venues for low-rent mass tourism, "places of inferiority which stood for everything that dominant social groups held to be tasteless, common and vulgar" (Urry 1990, 16).

The National Park Service (NPS) was created in 1916 to conserve majestic landscapes and to provide for their enjoyment by the public, a dual purpose that has struck many observers as oxymoronic, as tourist use may in fact compromise the natural attributes of place. Founding NPS Director Stephen Mather recognized the importance of park concessions to an enjoyable public experience, noting that "scenery is a hollow enjoyment to a tourist who sets out in the morning after an indigestible breakfast" (cited in National Park Service 2023). In fact, contrary to imagined "pure" conservation motives, the founding of the first US National Park, Yellowstone, in 1872 can be attributed to the influence of potential concessioners, executives from the Northern Pacific Railroad who anticipated a lucrative tourist trade.

Over the next century, the invention of the automobile, the development of the interstate highway system, and the increasing accessibility of air travel allowed a more-mobile-than-ever middle class to reach locations previously unthinkable for them. Tourism by car served audiences more interested in novelty, recreation, and experience than the cultural enlightenment promised in turn-of-the-twentieth-century park travels (Rothman 1998, 149; see also Runte 1987). For travelers of the 1920s, "travel was not designed to make them better, wiser, or more prepared; it merely restored them to their native condition, the way they had been before the rigors of urban civilized life wore them down" (Rothman 1998, 150). But the challenge to American values wrought by the advent of the Great Depression presented the parks with an opportunity to interpret, educate, and inspire, returning in some ways to the earlier Romantic impulse.

A growing national park system in the US, reached by newly available modes of transportation, became a site for personal expression for the striving professional and managerial middle class. But this did not please everyone. In the 1950s, members of the National Parks Association registered concern about the ways in which the public was using the parks. Paul Shepherd Jr. "bemoaned the commercialization and rapid pace of park visits, and felt that 'a larger majority of visitors [were] unaware of the peculiar meaning ... of the parks'" (Carr 2007, 54). The resurrection of the Romantic impulse promised to tame the threat of the damaging, novelty-mad American tourist, as a battle for the soul of the park traveler played out alongside the emergence of the era of the lifestyle, wherein "one expressed oneself more at leisure than at work; by one's hobbies, one's possessions, one's tastes" (Feifer 1985, 224). Conservation-oriented wilderness sites, with limited tourist service infrastructure, are and were not fancy, but just exclusive enough to provide refuge from the tacky, the vulgar, the mass, and the inferior, even as tourism literally consumes the lands, threatening to deplete or exhaust that which makes the destination striking or significant (Urry 1995, 2).

As more people traveled, the mobility of the upper classes provided them with cultural capital that helped them to evaluate and distinguish between different environments and to register as important their impressions of which ones had value and which were to be despised (Urry 1995, 175). In the present day, Americans with high cultural capital tend to understand sophistication through cosmopolitanism, "as having had a chance to travel, to learn languages, to discover various culinary traditions, and, more generally, to widen one's horizons—which goes along with the quest for self-actualization" (Lamont 1992, 107). Vacation destinations and the eating experiences that accompany them signal refinement in this context, making national parks fascinating sites for struggles of cosmopolitan self-making.

Getting to Cosmopolitanism

Since their founding, national parks have been places for introspection, communing with nature, experiencing the sublime, finding one's authentic self, and also virtue-signaling. From the start, national parks were pitched toward those who could afford the time and substantial expense involved in traveling to remote locations—those with capital, economic and cultural. Victor and Edith Turner asserted that "modern pilgrimages may be read as 'meta-social commentaries' on the troubles of the epoch and a search for vanishing virtues" (1978, 38). In the late 1800s, wealthy travelers made exhausting and expensive pilgrimages to the parks to escape the dirty and overcrowded industrial cities of the east. The romantic ethos of national parks prioritized leaving behind the mundanity of daily life, instead traveling widely to commune with nature, to experience solitude and awe-inspiring vistas, and thus to reflect and change one's life for the better.

Perhaps not much has changed in the last 150 years since Yellowstone National Park was founded. Much as spas and springs served as resorts for the wealthy, for whom a medical motive for travel—taking the therapeutic waters—excused the traveler temporarily from the demands of the Protestant work ethic (Bartlett 1985), today's national parks may also serve not only an aesthetic problem for visitors but also a moral one. In the overworked US, where 52% of workers left vacation days on the table in 2017 (US Travel Association 2018), we still frown to some extent on vacations. But there is redemptive value in communing with nature, hiking, or achieving spiritual rejuvenation in the outdoors, so the parks can solve a moral problem for visitors. National park travel allows the leisure of visitors to pass as a quest for health, thus making their recreation both socially acceptable and morally satisfactory (Bartlett 1985, 114). The food experiences in the parks bear some examination, then, as they do not always serve this mission well.

US national parks remain attractive to travelers who gain distinction through turning away from hyper-connected, media-saturated reality and rampant consumption of a never-ending supply of goods; these travelers instead seek "nature," authenticity, and solitude as they travel slow, basking in the transformation offered by what has come to be known as the "experience economy." Today, highly mobile travelers from affluent parts of the world travel far, "often to take largely inward journeys: to practice 'simplicity' and 'slowness' and experience 'authenticity'" (Howard 2012, 12). But the global flows of information enabled by transportation networks and communication technologies mean that cosmopolitanism isn't just something that rich people do; rather, it's a cultural repertoire, available to many people across class and cultural boundaries.

Today, national parks are viable destinations for slow travel, which has emerged as a preferred status signifier. Slow travel exemplifies resistance to the artificial pleasures lamented by Olmsted, eschewing the fast, cheap, and out of control in favor of something quieter, smaller, and more special. Many elements of the parks lend themselves to slow travel, as their often-remote locations mean that they take time and effort to get to, and they're off the grid, lacking televisions, strong Wi-Fi, and sometimes even cell service. Reservations are harder to get at many park lodges and campsites than the hottest new urban eatery, their elusiveness and exclusivity adding to their appeal. These features, along with their magnificent scenery, make park travel particularly appealing for tasteful travelers who have been cultivated to make distinctions and who enjoy communing with nature and slowing down, and reaping the social value of doing these activities. But these travelers find a food scene that is not generally as inspirational as the scenery. The food available in national parks has tended to emphasize neither culinary heritage, organic principles, nor the sensory embodiment of the journey, all cosmopolitan-friendly features of slow travel (Fullagar, Wilson, and Markwell 2012, 4), but this is changing, as cosmopolitan values have started to inflect the future of food in the parks.

The national parks are well set up for slow travel, as a great number of them are remote or not easy to get to. In most, there's enough infrastructure that the traveler is never really required to "rough it," but there's enough distance from luxurious creature comforts (gourmet food, high thread-count sheets, light-speed Wi-Fi) that the travel itself reflects something importantly "alternative" about the traveler, enhancing their cultural capital (Munt 1994, 108). I'm particularly interested in the park food experiences of tastemakers and cultural intermediaries, for whom authentic experiences take precedence over experiences that require great amounts of economic capital (which they often do not have). Food is a way of communicating social identity, and through their cosmopolitan choices and attitudes, eaters can maintain distinctiveness from what they imagine as a parochial Other; however, they can also connect across differences.

Foodies are known to be anxious about consumerism and to find consumption morally freighted (de Solier 2013, 80). Educated foodies participate in cosmopolitan discourses when they imagine themselves as culturalists—people who self-make through experiences and the cultivation of knowledge—rather than materialists, who accumulate things; and they particularly love food because they can enjoy material culture without the anxiety-producing accumulation it usually entails (de Solier 2013, 81). This attitude places certain pressures on food experiences during national park travel that the NPS, concessioners, and enterprising locals have been shifting toward providing.

Although a small number of people live and work in the national parks, for most people, time spent there is not everyday life. Because it's vacation, there's no cooking to be done, except perhaps for cooking while camping out, in which case the contextual limitations prevent *real* cooking. Fire bans in many of the parks and concerns about animal and camper safety have produced a situation in which camp cooking is often quite limited or outright nonexistent; thus, there is no showcase for the morally precious labor of food production. This situation puts more pressure on national park restaurants, which are then (inevitably) even more disappointing, because they present so few opportunities for the kind of romantic consumption— that is, consumption that is "imaginative, remote from experience, visionary, and preferring grandeur or passion or irregular beauty" (Campbell 1987, 1)—that cosmopolitanism requires. The ethics and aesthetics of omnivorous cosmopolitans, which involve openness to the high and low, and to connecting across difference, but disregard for the facile—"easy, shallow, cheap, easily decoded and culturally undemanding" (Bourdieu 1984, 486)—are put to the test in the eating environments of national parks.

For those who can make the journey, parks offer limited opportunities for virtuous consumption alongside like-minded others. In the context in which "patterns of consumption and practices of the self such as eating and table manners mark class differences on to the body" (Bell and Valentine 1997, 23), eating and taste in national parks communicate about the status of the traveler. If everyone can go to the parks, how might one be set apart? As traditional markers of cultural legitimacy are shifting (Bellavance 2008, 190) and socioeconomic inequality rises in the late twentieth-century US, taste both reflects and shapes these inequalities; per Juliet Schor, "taste ceases to be a personally and socially innocent category" (1998, 29). The era of social media has made formerly invisible experiences like dining, leisure activities, and tourism effectively positional goods. Mass-produced goods have lost their ability to individuate and differentiate when everyone can buy them, but signature food experiences allow for a customized dimension that helps strivers keep up with upscale groups (Schor 1998, 76).

Food, then, is important in national park travel. Food consumption in tourism is both essential and symbolic: "Since food also serves as a 'social marker' which identifies one's group, social status is one of the pervasive factors affecting the types and quantity of foods eaten and the perceived meanings of foods" (Mak et al. 2012, 932). Food is a positional good that facilitates social differentiation, connecting people to others who consume in similar ways (Urry 1995, 131). Eating in the parks provides, at least in principle, an opportunity for the kinds of reflection, interpretation, and openness that are central to cosmopolitan discourses, and which align handsomely with Olmsted's emphasis on contemplative recreation in natural spaces.

What Is Cosmopolitanism?

Cosmopolitanism is a concept that has been embraced for its socially and political transformative potential (Rovisco and Nowicka 2011, 1). It's not just an abstract concept but is grounded in empirical reality and material practices—in what people *think* and *do*. Research on cosmopolitanism interrogates "negotiations of difference via empathetic engagements with other cultures and value-systems, issues of self-transformation, and mobilities of various kinds" (Rovisco and Nowicka 2011, 2). My work is not trying to ask who are cosmopolitan visitors to national parks, how many of them are there, what is their racial/ethnic profile and socioeconomic status, and what are they eating, but rather invokes cosmopolitanism as an analytical tool: it allows me to investigate how cosmopolitanism inflects taste and foodways in the parks, thus shaping the experiences of a range of groups and individuals, one bite at a time.

Food choices and attitudes display wealth and social standing, and food tourism allows people to demonstrate individualism, affluence, and distinction (Harrington and Ottenbacher 2010, 19). Simultaneously a vehicle for expressing cultural identity and a simple form of sustenance, food "is an ideal vehicle for studying the meanings of cosmopolitanism in everyday life" (Cappeliez and Johnson 2013, 435). Cosmopolitanism is place-based, as it relates to consumption across real and symbolic cultural borders and boundaries. As I have traveled through US national parks, I have thought about how the people who visit these spaces have shaped the food system in the parks and how the food system has shaped local cultures and landscapes alike. I have also been struck by the emergence, however scattered and inchoate, of the kinds of authentic eating experiences sought by today's romantic consumers—omnivorous citizens of the world eager to engage in global conversations, seeking mutual respect across differences (Appiah 2007)—into these highly industrialized food systems, and I'm heartened by the possibilities they present. How do the food experiences in national parks, otherwise ideal spaces for virtuous travel, constrain and provide opportunities for the expression of cosmopolitan identity? To answer this, I need to provide a framework for understanding what cosmopolitanism is and what it isn't.

Connoisseurs and Oppression: One View of Cosmopolitanism

In the last three decades, much sociological research has been done on cosmopolitanism. Some scholars have imagined cosmopolitanism as a more-or-less stable identity trait and have accordingly written about "cosmopolitans" as if they were a coherent group marked by privilege and mobility. For instance, Skey defines cosmopolitans as "members of more 'elite' groups that travel widely, engage with a variety of cultural groups and practices, work or live abroad or represent the vanguard of 'global' social, political or cultural movements" (2012, 474). The "universal citizenship" of cosmopolitanism allows travelers to function across cultural lines, demonstrate "a flexible intellectual aesthetic openness towards divergent cultural experiences, towards contrasts (difference) rather than uniformity (sameness), and towards the allure of elsewhere and Otherness" (Salazar 2010, 78). Aspirational and educated members of the middle class, cosmopolitans, from this perspective, are known to be particularly interested in things like camping, hiking, yoga, and theatre- and museum-going as forms of recreation (Savage et al. 1992, 108).

Cosmopolitans are concerned with health and fitness, and these educated members of the professional and managerial middle class tend to be outdoorsy, enjoy spending time in nature, and often belong to environmental organizations (Urry 1995, 226). For health-focused cosmopolitans, the body and the environment can take on spiritual significance as parts of nature as a whole; in this context, the outdoors provides a site for healthy recreation and symbolizes society's failure to nurture nature (Urry 1995, 226). According to Lamont, "vacation destinations and eating habits are used as signals of refinement" (1992, 107) for cosmopolitans, who define sophistication as being able to widen one's horizons in a quest for self-actualization. Spaces dedicated to conservation and enjoyment, like national parks, then become particularly interesting spaces in which cosmopolitan body projects are carried out. Thus, industrialized park food is a pain point for privileged, highly mobile consumers.

This perspective on cosmopolitanism suggests that there are many motives for and expressions of cosmopolitan good taste, including wonder, curiosity, authenticity, distinction, trophy-taking, status, and national/cultural identity, and these motives play out in park tourism in interesting ways. In this view, highly mobile cosmopolitans, unbound by location, see in wilderness landscapes an opportunity to deploy their well-honed faculties for aesthetic judgment and appreciation. But the wonder and curiosity that cosmopolitans bring to these wild spaces are not innocent. "The production of wonder ... is a calculated rhetorical strategy, the evocation of an aesthetic response in the service of a legitimate process" (Greenblatt 1991, 73–74). Cosmopolitan travelers, urged to "take only memories," value experiences (and their representation via photographs) that create wonder and spur curiosity, not the trinkets, souvenirs, or postcards of the less sophisticated or foreign "trophy-taking" tourist (Machlis, Field, and Van Every 1984). The landscape itself is the object to be collected, endowing travelers with "the ownership of experience, and so with the experience of ownership" (Benedict 2002, 17–18).

The cosmopolitan travel experience has deep roots in what Shaffer has described as the "elite cultural tradition of secular pilgrimage" (Shaffer 2001, 4). White, well-off, US-born Americans were encouraged by an emerging tourist industry in the first 60 years of the national parks to virtuously consume the nation through tourism, thus becoming better Americans (Shaffer 2001, 4). "Tourism ... promised to reconcile this national mythology, which celebrated nature, democracy, and liberty, with the realities of an urban-industrial nation-state dependent on extraction, consumption, and hierarchy" (Shaffer 2001, 5). Their mobile citizenship emphasized a commercial form of citizenship born out of modern consumer capitalism:

> a one-of-a-kind personal experience as a mass-produced phenomenon. Because tourism trafficked in the sale of experiences and spectacle rather than objects, it could promise a singular, personal, adventure for each person. In this way tourism distanced itself from the process of commodification.
>
> *(Shaffer 2001, 264)*

With the birth of the automobile and critical infrastructure developments, tourism became democratized (at least to the middle classes) in the first half of the 1900s.

Cosmopolitan taste is neither the taste of the peasant nor that of the fat cat. Instead, it occupies a kind of middle space between the wealthy and the working classes associated by Bourdieu with teachers, whose ascetic consumption is born of their surplus of cultural capital and surfeit of economic capital. Cosmopolitans pursue "originality at the lowest economic cost and go in for exoticism ... and culinary populism," defining their taste against the vulgar forms of consumption at either end of the economic spectrum (Bourdieu 1984, 185). Eating a meal imagined as "a social ceremony, an affirmation of ethical tone and aesthetic refinement" (Bourdieu 1984, 196) allows bourgeois citizens to aestheticize the deferred and the difficult, thereby capitalizing on their taste preferences for the authentic, the natural, and the real. Cosmopolitans are middle class, if professionally and managerially so, and they "graze" omnivorously across cultural forms (Sullivan and Katz-Gero 2007; Bennett et al. 2009, 18); this openness itself confers distinction, as eclecticism is hailed as "the new 'cool'" (Bellavance 2008).

But this cosmopolitan hunger for authenticity is in some ways a strange impulse. Kirshenblatt-Gimblett wonders:

> Why, if we do not debate the authenticity of the toast and coffee of our daily breakfast, do we become anxious about the authenticity of an ethnic restaurant or gastronomic travel experience? Restaurants are prime sites of designed experiences, collaboratively produced. Our preoccupation with their authenticity goes to the heart of the concept of culinary tourism: namely, how self-consciousness

arises from encounters with the unfamiliar and challenges what we know—or think we know—about what is before us.

(2004, xii)

Educated members of the middle class, cosmopolitans aren't terribly interested in cultural forms that are purely abstract; instead, they "seek to evoke, even possess, the 'ordinary' and the 'real' through the act of cultural appropriation itself" (Bennett et al. 2009, 71). They relish opportunities to apply reflective judgment across different cultural forms in a variety of contexts (Bennett et al. 2009, 71), and restaurant dining, with its relatively low economic bar to entry, is a particularly attractive arena for the exhibition of their knowledge and taste. Cosmopolitan versatility of taste provides the upwardly mobile middle classes an alibi—a safe distance from accusations of elitism—and an endorsement of "the consumerist imperative to try everything in search of new gratifications" (Bennett et al. 2009, 255).

This view of cosmopolitanism lends itself to a critique of those privileged, educated, highly mobile people who have capital and who know how to use it, because these people use capital in ways that, unwittingly or not, oppress others. Scholars who adopt this perspective view cosmopolitanism warily, as an expression of the cultural logic of global capitalism, exploiting Others and selling difference (Žižek 1997; Harvey 2009).

There is an inherent racism in the expression of cosmopolitanism from this perspective. For instance, Savage, Wright, and Gayo-Cal point out the white privilege of cosmopolitan practices: while white people may wish to participate in forms of cultural consumption that allow them to learn more about the Other, what they want is an Other that is not too threatening, and that flatters their (white, educated, middle class) worldview (2010, 612). Similarly, Rodriguez laments Anglo white cosmopolitan desire for transcendence and redemption through contact with Indigenous subjects imagined as authentic and pristine (Rodriguez 2003). The experiences that widen the horizons of white cosmopolitan travelers hinge on their difference, enabled by an objectifying, instrumentalist, ethically problematic view of the lands and the local (especially Indigenous) people as simply there to be looked at or experienced (MacCannell 2011, 21).

Where cosmopolitanism inflects culinary tourism, Germann Molz recognizes a performance of white, Western sense of adventure, adaptability, and openness that reinforces cultural norms and enhances status, rather than a desire for genuinely connecting with another culture (2007, 77). Cosmopolitanism contains an interesting paradox, with ideals of inclusiveness across differences chafing against cosmopolitanism as an indicator of distinction (Germann Molz 2011). Furthermore, critics of cosmopolitanism pose pressing questions about how consumption translates to other commitments, urging us to consider whether cosmopolitan consumption changes people's values and habits, and their assumptions, and makes them commit to riskier forms of political activism (Germann Molz 2011, 48; see also Hall 2012; Markwell, Fullagar, and Wilson 2012; Laing and Frost 2015).

Pragmatic Connectors: Cosmopolitanism and Transformation

A different way of seeing cosmopolitanism, instead of as a stable identity trait of privileged, outdoorsy travel-loving foodies, is to understand it as a more flexible opportunity. It's not just elites who desire new and different foods and want to feel comfortable and knowledgeable doing so (Warde 2018)—these attitudes and practices exist across the class and status spectrum. Cappeliez and Johnston see cosmopolitanism as a cultural repertoire and recognize that there are different modes of cosmopolitan consumption. While "connoisseur mode" exemplifies privilege and agency, may involve some degree of knowledge showboating, and endorses and reinforces taste hierarchies (with guess who on top!), a "pragmatic mode" that centers social connections and lived experiences is open to people across a range of socioeconomic positions, even those with relatively limited geographic mobility, and results in genuine bonds that shift values (Cappeliez and Johnston 2013, 447), and so it's worth looking at the transformative power of this mode.

Similarly, Skrbis and Woodward helpfully posit cosmopolitanism as "an increasingly prominent, available cultural discourse – and ideal" rather than an identity that is fixed to one's social position (2007, 735); it's something people draw on ambivalently, "a tool for negotiation of life chances in an increasingly interconnected and open world" (Skrbis and Woodward 2007, 746). Cosmopolitan imagination and consumption are shaped by race, class, and gender (Nava 2002). These identity vectors influence people's consumption and production practices, giving some privilege and easy access, and others greater challenges to negotiating cosmopolitan desires. Cosmopolitanism correlates positively with ethical consumption behavior and demonstrated concern for the environment (Pecoraro and Uusitalo 2014; Grinstein and Riefler 2015).

Although critics of cosmopolitanism offer important food for thought, other scholars argue that cosmopolitanism genuinely creates new modes of cross-cultural interaction, interconnectedness, and understanding (Appadurai 1996; Beck 2000), especially around issues like environmental degradation that endanger people across borders (Beck 2006). To say that all cosmopolitanism is complicity in oppression is a hasty overdetermination that denies the plurality of cosmopolitanism. Some scholars articulate the possibility of cosmopolitan reflexiveness—not simply a pantomime of appreciation, but "a deeper and more skilled engagement with otherness" that "shows some desire and willingness to be challenged and learn from different cultural experiences, and [...] some implicit value preference for the explicit de-hierarchization of cultural on political or ethical grounds" (Skrbis and Woodward 2011, 61). It is this willingness "to step outside stable, privileged and established power categories of selfhood" (Skrbis and Woodward 2011, 61) that marks the transformative potential of cosmopolitanism.

Skrbis and Woodward encourage researchers to look at how cosmopolitans' contact with others in consumptive, commodified contexts like tourism and food might lead to more critical forms of cosmopolitanism (2011, 64). There doesn't just have to be a thin or elitist cosmopolitanism that is a reductive form of worldliness—rather, there can be critical cosmopolitanism—a "worldly sensibility from below" (Kurasawa 2011, 281) that critiques liberal multicultural diversity and that radically decenters one's own position and worldviews and involves deep critique and interrogation resulting in a "normative commitment to the moral equality of all human beings" (Kurasawa 2011, 282). Cosmopolitan discourses contain both an aesthetic element, involving openness and curiosity toward difference, and a moral side, involving concern for others, not just members of one's own tribe (Skrbis, Kendall, and Woodward 2004). This would seem to bode well for actual transformational social change.

There are different ways of expressing cosmopolitanism with regard to food: whereas previously distinction might be expressed through connoisseurship of exotic foods, new forms of cosmopolitanism hinge on valorizing the local (Parasecoli 2017, np). Ho identifies a cosmopolitan locavorism, as distinct from "defensive localism, the cultural consumption of globetrotting elites, or an urban fascination with rural lifestyles. It simultaneously signifies trans-local connections and human-land bonds, mobilizing a cultural critique of local neoliberal governmentality" (Ho 2020, 137). Instead of a cartoonish ideology held by wealthy, Western elite world travelers, cosmopolitanism can inflect localist food activism in striking and significant ways. Instead of an oversimplified, over-romanticized view of "the local as unquestionably ethical, progressive, and high-quality, while the non-local as polluting, untrustworthy, and poor-quality," cosmopolitan locavorism "emphasizes connection and inclusion as opposed to separation and exclusion" (Ho 2020, 138). Cosmopolitan locavore values infuse the bioregionalist and Indigenous-centered national park food initiatives I address later in the book. Because of their special status of places of conservation, interpretation, and enjoyment, national park food spaces can offer the kinds of explicitly taught cosmopolitan cultural scripts that can engender transformative implicit cultural learning, posing a true challenge to existing hierarchies of power and value (Høy-Petersen 2021, 1207). Local and ethical food movements have demonstrated that "ideals of taste, aesthetics, and ethics are not incompatible" (Emontspool and Georgi 2017, np). National parks provide opportunities to recreate and consume in ecologically sustainable ways. Against this backdrop, the fulfilment of cosmopolitan tastes can be transformative.

The question of food tourism in national parks is an interesting one. The parks tend to be associated with wilderness, which is by definition not a space for agriculture or population centers, so there would seem to be some challenges to the most appealing features of cosmopolitan food travel. Accessibility, seasonality, and expense are difficult problems for small-scale, nonindustrial food providers, but cosmopolitan tastes are starting to shift even the industrial concessions toward the local and seasonal. Critics have advocated for forms of national park tourism that give the tourist a distinctive sense of the destination and serve the local population (Sax 1980, 107), a mission aligned with the agenda and expectations of today's cosmopolitan travelers, but most of the food opportunities currently fall short of the mark. Wouldn't the cultivation that Olmsted sought for Yosemite be better accomplished by something that occasioned the use of the contemplative faculties, instead of just feeding the gullet?

In a hypermobile world, where everything we want is a keystroke away, food experiences that are slow and immobile are highly appealing. Pragmatically cosmopolitan travelers are drawn by the "specificity of experiencing it on the spot, in relation to season, ripeness, freshness, perishability, and total world of which it is a part—that requires that you go there" (Kirshenblatt-Gimblett 2004, xiv). They build on touristic terroir—"the unique combination of the physical, cultural and natural environment gives each region its distinctive touristic appeal" (Hall, Mitchell, and Sharples 2003, 34)—to animate food products and experiences in ways that differentiate them from the quotidian, the functional, and the standardized.

Although they are still, at this time, outnumbered by a prominent parade of heat lamp-warmed, industrial chicken sandwiches in cafeterias lacking in charm, there *are* eating opportunities in the parks that contribute to the parks' interpretive mission. Opportunities for cooking classes where visitors learn from Indigenous locals; farm tours; a shared meal in someone's home; and the chest-thumping allure of the seasonal, the artisanal, terroir, foodshed, and the like are appearing on the national parks scene, activating visitors' talents of perceiving, evaluating, and distinguishing, and giving a chance for genuine connection. When these kinds of prized food experiences do exist in the parks—the fancy restaurant with the gob-smacking views that sources locally, the kitschy cowboy cookout that occasions reflection on the performance of authenticity, the pick-your-own orchard where one can pluck heirloom fruits right from the tree and post to Instagram in one fell swoop—their emergence is largely due to the demand of cosmopolitans.

To understand how and why cosmopolitan tastes are transforming park foodways, it is necessary to look backward to consider the evolution of the parks as foodscapes. A different set of desires animated the development of the industrial concession system that dominates today's parks—and these appear very much in tension with both the Romantic beginnings and cosmopolitan futures of US national park foodways.

References

Appadurai, Arjun. 1996. *Modernity at Large: Cultural Dimensions of Globalization.* Minneapolis: University of Minnesota Press.

Appiah, Kwame Anthony. 2007. *Cosmopolitanism: Ethics in a World of Strangers.* New York: W.W. Norton.

Bartlett, Richard A. 1985. *Yellowstone: A Wilderness Besieged.* Tucson: University of Arizona Press.

Beck, Ulrich. 2000. "The Cosmopolitan Perspective: Sociology of the Second Age of Modernity." *British Journal of Sociology* 51: 79–105.

Beck, Ulrich. 2006. *The Cosmopolitan Vision.* Cambridge: Polity.

Bell, David and Gill Valentine. 1997. *Consuming Geographies: We Are Where We Eat*. New York: Routledge.

Bellavance, Guy. 2008. "Where's High? Who's Low? What's New? Classification and Stratification Inside Cultural 'Repertoires.'" *Poetics* 36 (2–3): 189–216.

Benedict, Barbara. 2002. *Curiosity: A Cultural History of Early Modern Inquiry*. Chicago, IL: University of Chicago Press.

Bennett, Tony, Mike Savage, Elizabeth Silva, Alan Warde, Modesto Gayo-Cal, and David Wright. 2009. *Culture, Class, Distinction*. New York: Routledge.

Bourdieu, Pierre. 1984. *Distinction: A Social Critique of the Judgement of Taste*. Translated by Richard Nice. New York: Routledge and Kegan Paul.

Brillat Savarin, Jean Anthelme. 1825. *The Physiology of Taste; or, Transcendental Gastronomy*. Trans. Fayette Robinson. https://www.gutenberg.org/cache/epub/5434/pg5434.txt.

Campbell, Colin. 1987. *The Romantic Ethic and the Spirit of Modern Consumerism*. Oxford: Blackwell.

Cappeliez, Sarah and Josée Johnston. 2013. "From Meat and Potatoes to 'Real-Deal' Rotis: Exploring Everyday Culinary Cosmopolitanism." *Poetics* 41 (5): 433–55. https://doi.org/10.1016/j.poetic.2013.06.002.

Carr, Ethan. 2007. *Mission 66: Modernism and the National Park Dilemma*. Amherst: University of Massachusetts Press.

de Solier, Isabelle. 2013. *Food and the Self: Consumption, Production and Material Culture*. London: Bloomsbury.

Emontspool, Julie and Carina Georgi. 2017. "A Cosmopolitan Return to Nature: How Combining Aesthetization and Moralization Processes Expresses Distinction in Food Consumption." *Consumption, Markets and Culture* 20 (4): 306–28. https://doi.org/10.1080/10253866.2016.1238823.

Feifer, Maxine. 1985. *Tourism in History: From Imperial Rome to the Present*. New York: Stein and Day.

Fisher, Irving D. 1986. *Frederick Law Olmsted and the City Planning Movement in the United States*. Ann Arbor: UMI Research Press.

Fullagar, Simone, Erica Wilson, and Kevin Markwell. 2012. "Starting Slow: Thinking Through Slow Mobilities and Experiences." In *Slow Tourism: Experiences and Mobilities*, edited by Simone Fullagar, Kevin Markwell, and Erica Wilson, 1–8. Bristol: Channel View.

Germann Molz, Jennie. 2007. "Eating Difference: The Cosmopolitan Mobilities of Culinary Tourism." *Space and Culture* 10 (1): 77–93.

Germann Molz, Jennie. 2011. "Cosmopolitanism and Consumption." In *The Ashgate Research Companion to Cosmopolitanism*, edited by Maria Rovisco and Magdalena Nowicka, 33–52. Burlington, VT: Ashgate.

Gigante, Denise. 2007. "Romantic Gastronomy: An Introduction." In *Romantic Gastronomies*, edited by Denise Gigante. https://romantic-circles.org/praxis/gastronomy/gigante/gigante_essay.html.

Greenblatt, Stephen. 1991. *Marvelous Possessions: The Wonder of the New World*. Chicago, IL: University of Chicago Press.

Grinstein, Amir and Petra Riefler. 2015. "Citizens of the (Green) World? Cosmopolitan Orientation and Sustainability." *Journal of International Business Studies* 46: 694–714.

Hall, C. Michael. 2012. "The Contradictions and Paradoxes of Slow Food: Environmental Change, Sustainability and the Conservation of Taste." In *Slow Tourism: Experiences and Mobilities*, edited by Simone Fullagar, Kevin Markwell, and Erica Wilson, 53–68. Bristol: Channel View.

Hall, C. Michael, Richard Mitchell, and Liz Sharples. 2003. "Consuming Places: The Role of Food, Wine and Tourism in Regional Development." In *Food Tourism Around the World: Development, Management and Markets*, edited by C. Michael Hall, Liz Sharples, Richard Mitchell, Niki Macionis, and Brock Cambourne, 25–59. Burlington, MA: Butterworth Heinemann.

Harrington, Robert J. and Michael C. Ottenbacher. 2010. "Culinary Tourism—A Case Study of the Gastronomic Capital." *Journal of Culinary Science and Technology* 8 (1): 14–32.

Harvey, David. 2009. *Cosmopolitanism and the Geographies of Freedom*. New York: Columbia University Press.

History of the Train. 2022. "Grand Canyon Railway and Hotel." Accessed August 8, 2022. https://www.thetrain.com/the-train/history/.

Ho, Hao-Tzu. 2020. "Cosmopolitan Locavorism: Global Local-Food Movements in Postcolonial Hong Kong." *Food, Culture and Society* 23 (2): 137–54. https://doi.org/10.1080/15528014.2019.1682886.

Howard, Christopher. 2012. "Speeding Up and Slowing Down: Pilgrimage and Slow Travel Through Time." In *Slow Tourism: Experiences and Mobilities*, edited by Simone Fullagar, Kevin Markwell, and Erica Wilson, 11–24. Bristol: Channel View.

Høy-Petersen, Nina. 2021. "Civility and Rejection: The Contextuality of Cosmopolitan and Racist Behaviours." *Sociology* 55 (6): 1191–210. https://doi.org/10.1177/00380385211011570.

Kirshenblatt-Gimblett, Barbara. 2004. "Foreword." In *Culinary Tourism*, edited by Lucy M. Long, xi–xiv. Lexington: University Press of Kentucky.

Kurasawa, Fuyuki. 2011. "Critical Cosmopolitanism." In *The Ashgate Research Companion to Cosmopolitanism*, edited by Maria Rovisco and Magdalena Nowicka, 279–93. Burlington, VT: Ashgate.

Laing, Jennifer and Warwick Frost. 2015. "The New Food Explorer: Beyond the Experience Economy." In *The Future of Food Tourism: Foodies, Experiences, Exclusivity, Visions and Political Capital*, edited by Ian Yeoman, Una McMahon-Beattie, Kevin Fields, Julia N. Albrecht, and Kevin Meethan, 177–93. Bristol: Channel View.

Lamont, Michèle. 1992. *Money, Morals and Manners: The Culture of the French and American Upper-Middle Class*. Chicago, IL: University of Chicago Press.

Lobel, Cindy R. 2014. *Urban Appetites: Food and Culture in 19th Century New York*. Chicago, IL: University of Chicago Press.

MacCannell, Dean. 2011. *The Ethics of Sightseeing*. Los Angeles: University of California Press.

Machlis, Gary E., Donald R. Field, and Mark E. Van Every. 1984. "A Sociological Look at the Japanese Tourist." In *On Interpretation: Sociology for Interpreters of Natural and Cultural History*, edited by Gary E. Machlis and Donald R. Field, 77–93. Corvallis: Oregon State University Press.

Mak, Athena H.N., Margaret Lumbers, Anita Eves, and Richard C.Y. Chang. 2012. "Factors Influencing Tourist Food Consumption." *International Journal of Hospitality Management* 31 (3): 928–36.

Markwell, Kevin, Simone Fullagar, and Erica Wilson. 2012. "Reflecting Upon Slow Travel and Tourism Experiences." In *Slow Tourism: Experiences and Mobilities*, edited by Simone Fullagar, Kevin Markwell, and Erica Wilson, 227–33. Bristol: Channel View.

Meeker, Joseph W. 1973. "Red, White, and Black in the National Parks." *The North American Review* 258 (3): 3–7.

Munt, Ian. 1994. "The 'Other' Postmodern Tourism: Culture, Travel, and the New Middle Classes." *Theory, Culture and Society* 11 (3): 101–23.

National Park Service. 2023. "Commercial Services Program." Accessed January 7, 2023. https://www.nps.gov/orgs/csp/index.htm?.

Nava, Mica. 2002. "Cosmopolitan Modernity: Everyday Imaginaries and the Register of Difference." *Theory, Culture and Society* 19 (1–2): 81–99.

Olmsted, Frederick Law. [1865] 1994. "The Yosemite Valley and the Mariposa Big Tree Grove: Olmstead Report on Management of Yosemite, 1865." In *America's National Park System: The Critical Documents*, edited by Lary M. Dilsaver, 6–19. Lanham, MD: Rowman and Littlefield. https://www.nps.gov/parkhistory/online_books/anps/anps_1b.htm.

Parasecoli, Fabio. 2017. "Cosmopolitan Foodies and Local Food." *The Huffington Post*, December 4. https://www.huffpost.com/entry/cosmopolitan-foodies-and-local-food_b_5a2534cee4b04dacbc9bd8e6.

Pecoraro, Maria and Outi Uusitalo. 2013. "Conflicting Values of Ethical Consumption in Diverse Worlds - A Cultural Approach." *Journal of Consumer Culture*, 14, 45–65. https://doi.org/10.1177/1469540513485273.

Rodriguez, Sylvia. 2003. "Tourism, Difference, and Power in the Borderlands." In *The Culture of Tourism, the Tourism of Culture: Selling the Past to the Present in the American Southwest*, edited by Hal K. Rothman, 185–205. Albuquerque: University of New Mexico Press.

Rothman, Hal K. 1998. *Devil's Bargains: Tourism in the 20th Century American West*. Lawrence: University of Kansas Press.

Rovisco, Maria and Magdalena Nowicka. 2011. "Introduction." In *The Ashgate Research Companion to Cosmopolitanism*, edited by Maria Rovisco and Magdalena Nowicka, 1–14. Burlington, VT: Ashgate.

Runte, Alfred. 1987. *National Parks: The American Experience*. Lincoln: University of Nebraska Press.

Salazar, Noel B. 2010. *Envisioning Eden: Mobilizing Imaginaries in Tourism and Beyond*. New York: Berghahn.

Savage, Michael, James Barlow, Peter Dickens, and Tom Fielding. 1992. *Property, Bureaucracy and Culture: Middle Class Formation in Contemporary Britain*. London: Routledge.

Savage, Mike, David Wright, and Modesto Gayo-Cal. 2010. "Cosmopolitan Nationalism and the Cultural Reach of the White British." *Nations and Nationalism* 16 (4): 598–615. https://doi.org/10.1111/j.1469-8129.2010.00449.x.

Sax, Joseph L. 1980. *Mountains Without Handrails: Reflections on the National Parks*. Ann Arbor: University of Michigan Press.

Schor, Juliet. 1998. *The Overspent American: Upscaling, Downshifting and the New Consumer*. New York: Basic Books.

Shaffer, Marguerite. 2001. *See America First: Tourism and National Identity, 1880–1940*, Washington, DC: Smithsonian Institution.

Skey, Michael. 2012. "We Need to Talk about Cosmopolitanism: The Challenge of Studying Openness Towards Other People." *Cultural Sociology* 6 (4): 471–87.

Skrbis, Zlatko, Gavin Kendall, and Ian Woodward. 2004. "Locating Cosmopolitanism: Between Humanist Ideal and Grounded Social Category." *Theory, Culture and Society* 21 (6): 115–36.

Skrbis, Zlatko and Ian Woodward. 2007. "The Ambivalence of Ordinary Cosmopolitanism: Investigating the Limits of Cosmopolitan Openness." *The Sociological Review* 55 (4): 730–47.

Skrbis, Zlatko and Ian Woodward. 2011. "Cosmopolitan Openness." In *The Ashgate Research Companion to Cosmopolitanism*, edited by Maria Rovisco and Magdalena Nowicka, 53–68. Burlington, VT: Ashgate.

Sullivan, Oriel and Tally Katz-Gero. 2007. "The Omnivorousness Thesis Revisited: Voracious Cultural Consumers." *European Sociological Review* 23 (2): 123–37.

Swiggett, Glen Levin. 1903. "What Is Romanticism?" *The Sewanee Review* 11 (2): 144–60. https://www.jstor.org/stable/27530553.

Turner, Victor Witter and Edith L.B. Turner. 1978. *Image and Pilgrimage in Christian Culture: Anthropological Perspectives*. New York: Columbia University Press.

Urry, John. 1990. *The Tourist Gaze: Leisure and Travel in Contemporary Societies*. London: Sage.

Urry, John. 1995. *Consuming Places*. London: Routledge.

US Travel Association. 2018. "State of American Vacation." Accessed May 1, 2019. https://www.ustravel.org/system/files/media_root/document/StateofAmericanVacation2018.pdf.

Warde, Alan. 2018. "Changing Tastes? The Evolution of Dining Out in England." *Gastronomica* 18 (4): 1–12.

Zaslowsky, Dyan and Tom H. Watkins. 1994. *These American Lands: Parks, Wilderness, and the Public Lands*. Washington, DC: Island Press.

Žižek, Slavoj. 1997. "Multiculturalism, or, the Cultural Logic of Multinational Capitalism." *New Left Review* 225, 28–52.

2

INDUSTRIAL FOOD IN THE WILDERNESS

Dining and Democracy

The founding impulses of the US national parks reflected Romantic notions of self-making, as almost a century after the American Revolution, post-revolution nationalism still gripped the country. In 1872, with the establishment of Yellowstone National Park, the US is widely thought to be the first nation to create a national park system,[1] a move that environmental historian and writer Wallace Stegner has referred to as "America's Best Idea." At a time when the young American nation sought distinction from Europe, the parks promised to showcase breathtaking landscapes whose vistas rivaled any of Europe's grand, human-built palaces, castles, or cathedrals. The parks thus provided US citizens with a much-needed rejoinder to persistent claims of European cultural superiority. Although set apart from Europe, certain Americans—the "white, property-owning men who served as the nation's prototypical citizens" (Klein 2020, 2)—imported, seemingly wholesale, from Europe a view of taste that was freighted with social and political significance. "At the same time, those excluded from this narrow conception of citizenship recognized in eating an accessible means of demonstrating their own sense of national belonging, as well as additional and, at times, explicitly oppositional aesthetic theories" (Klein 2020, 2). These conflicting taste aesthetics would come to find themselves played out in national park eating experiences for well over 100 years.

Informed by Romantic and Transcendental sensibilities popular among the wealthy Eastern Seaboard liberal elite, many well-resourced Americans believed that the individual imagination, properly inspired by nature, would propel individual rights and liberties. The country's glorious natural spaces, particularly in the West, reassured Americans of their status despite their lack of the palaces, cathedrals, museums, and castles of their European forebears. Taking succor in majestic American landscapes not only soothed cultural anxieties but also provided a space for citizens to resolve their conflicted feelings about the rapid industrialization

DOI: 10.4324/9781003455516-3

taking place in the East. They sought in the parks less pure wilderness itself, and more the vision of the sublime that had become well represented on postcards, in advertising, and in fine art and associated with status and sophistication (Schmitt 1969). Said landscape painter Thomas Cole,

> there are those who regret that with the improvements of cultivation the sublimity of the wilderness should pass away: for those scenes of solitude from which the hand of nature has never been lifted, affect the mind with a more deep toned emotion than aught which the hand of man has touched. Amid them the consequent associations are of God the creator—they are his undefiled works, and the mind is cast into the contemplation of eternal things.
>
> *(Cole 1836)*

Of course, the very "wilderness" so venerated by the Romantics and Transcendentalists in the early to mid-nineteenth century was itself a man-made invention, even if it was broadly understood as quite the opposite. In 1849, Thomas Carlyle, in a letter to Ralph Waldo Emerson, imagined settlers "steering over the Western Mountains to annihilate the jungle, and bring bacon and corn out of it for the Posterity of Adam!" (cited in Root and de Rochemont 1976, 189). Although wilderness was first imagined as in need of taming to make way for agriculture, the ecologically catastrophic Civil War challenged national identity; alongside the Industrial Revolution's rapacious appetite for natural resources and transformative notions of masculinity, the combination posed a stark threat to the Romantic wilderness concept. Industrialization changed how white, property-owning men conceived of themselves and was seen as a threat to white masculinity: "American men began to link their sense of themselves as men to their position in the volatile marketplace, to their economic success ... Now manhood had to be proved" (Kimmel 2008, 6). Witnessing the disastrous environmental impact of both the Civil War and the rapidly industrializing East, with the threat to masculinity a shadow issue simmering on the back burner and influenced by the 1864 publication of George Perkins Marsh's *Man and Nature*, a seminal work of US environmentalism, many turned from thoughts of how to conquer the wilderness, and instead began considering how to save it from the threat of industrialization.

The national parks created precisely for this purpose—to conserve and enshrine such wilderness—were in fact invented after centuries of American Indian wars invested in removing Indigenous peoples from the landscapes they had actively shaped for thousands of years (Solnit 2000; Williams 2000; Williams 2002; Baylor University 2011; Roos 2020). Wilderness, the idea of untamed, uninhabited lands, was a dream cultivated by the white and the wealthy. As William Cronon points out:

> The dream of an unworked natural landscape is very much the fantasy of people who have never themselves had to work the land to make a living – urban folk for whom food comes from a supermarket or a restaurant instead of a field, and

for whom the wooden houses in which they live and work apparently have no meaningful connection to the forests in which trees grow and die. Only people whose relation to the land was already alienated could hold up wilderness as a model for human life in nature, for the romantic ideology of wilderness leaves no place in which human beings can actually make their living from the land.

(1995, 42)

In the mid-nineteenth century, as conversations about conservation were intensifying, would-be park concessioners began to recognize the profits to be made from the sale of this romantic wilderness-infatuated flight from history. "The owners of the great machines of monopoly capital—the so-called means of production— were, with excellent reason, at the forefront of nature work because it was one of the means of production of race, gender and class" (Haraway 1984, 52). With the completion of the Transcontinental Railroad in 1869 came much speculation about the likely return on investment that the railroad companies could produce by providing tourists with the concessions that would allow them to commune with nature in newly conserved spaces. Railroads and independent entrepreneurs alike began to set up concessions—a railroad spur here, a tent camp, a primitive hotel, a bathhouse, or a saloon there—to ease the arduous journey through spaces of natural splendor, thereby producing the classed conditions of eventual cosmopolitanism.

Early Concessions and the Birth of Taste

After Yellowstone in 1872, other parks began to pop up around the country— Yosemite, Sequoia, and Mount Rainier all before 1900, as well as later-demoted Mackinac and Rock Creek, each with their own needs for concessions—alongside anxious chatter about the specter of "bad development." Coney Island and Niagara Falls loomed large in the elite imagination as cautionary tales about the market catering too democratically to mass tastes at sites of natural splendor. At both sites, development run amok detracted from the view: factories, mines, and mills lined the rim of Niagara Falls; and amusement parks, racetracks, hotels, and restaurants stood between visitors and the seashore at Coney Island. However, such overde- velopment did not deter throngs of humanity from visiting each site. By 1894, *The New York Times* was applauding efforts to bring a more "respectable" class of people to Coney's "Sodom-by-the-Sea," despite the chagrin of "'dive' keepers and liquor sellers" who catered to the rowdy throngs. (Likely to their relief, that change did not happen quickly: a 1957 article in the *Buffalo Evening News* magazine noted the still "lurid" reputation of Coney Island [Roosevelt 1957, 1].) And although long hailed as the country's finest natural wonder, by 1885, *The New York Times* decried Niagara Falls' "employment of tawdry sensational attractions," noting:

the increasing ugliness everywhere: the destruction of all vernal beauty and freshness; the crowding of unsightly structures for manufactures of various kinds around the very brink of the falls; the incessant hounding of travelers, and the enormous exactions of which they are the victims.

(1885, 11)

While Coney Island and Niagara Falls were being overrun by so-called "rogues and unscrupulous operators" (Zaslowsky and Watkins 1994, 15), concessions were cropping up in the nation's first national parks in the nearly five decades between the 1872 founding of Yellowstone and the passage of the 1916 Organic Act which founded the National Park Service (NPS). Yosemite National Park, in California, offers a particularly interesting concession history animated by concerns over taste and class. Entrepreneurs had arrived in the Yosemite Valley at least a decade before Congress turned the land over to California to preserve and manage in 1864, convinced of the tourist dollars that would eventually flow into the valley. "Local businessmen, curious hunters, and roaming photographers and artists found their way into the valley, but the expected crowds took years to materialize. A trip to Yosemite was expensive and difficult, requiring weeks of travel and dangerous risks" (National Park Service 2021). Nonetheless, as tourists started to make their way to the park—at first, travelers of means in covered wagons bringing their own supplies and, later, a wider economic swath of humanity via train and automobile, relying more heavily on park restaurants and lodges—concession offerings evolved rapidly.

In the early years, concessions tended to be extremely basic and might consist of a campsite or a tent hotel and a cook. Crude lodging facilities began to crop up in the late 1850s, yet by 1870, the Yosemite Valley still lacked the kinds of hospitality amenities that increasingly affluent visitors sought out and that would allow them to position themselves distinctively—hence the 1871 opening by John C. Smith of The Cosmopolitan Bathhouse and Saloon. "Known simply as the Cosmopolitan, Smith's establishment offered Yosemite visitors two prime amenities – hot or cold baths at any time of the day or night, plus a very well-stocked bar. (Smith's mint juleps were a favorite)" (Janiskee 2011; see also Green 1987). The elegant Cosmopolitan was a favorite of affluent travelers from the East, offering a level of amenity unavailable at the rustic inns they were staying in. The visitor register, signed by four US presidents and famed individuals including Rudyard Kipling and William Randolph Hearst, testified to not only the class of the visitors but also the relief The Cosmopolitan brought weary yet well-heeled travelers during its brief 13-year existence. The establishment's very name reads like the throwing down of a gauntlet around status and social class: to be cosmopolitan is to be sophisticated and refined (Lamont 1992), to be a citizen of the world rather than a parochial hayseed. This kind of claim-staking to exalted status reflected elite discourses about class, taste, democracy, and distinction that were shaping the imagination about the future of US national parks.

A casualty of the growing conviction that saloons should not operate outside of hotels, The Cosmopolitan—with its prosperous clientele, its promise of the finer things in life, and its name that bespoke bespokeness—ceased operation in 1884, several years before Yosemite was designated as a national park. It nevertheless served as a precursor to the exclusive, luxury lodges that would eventually populate the national parks of the western US. As trains began to reach the parks—the Grand Canyon in 1901, Yosemite in 1907, and Yellowstone in 1908—so did food begin to arrive by train; local foodways were quickly sidelined. Railroad freight and passenger services grew in tandem, increasing the variety of available foods, releasing people from the constraints of seasonal eating (Wallach 2013), and accelerating the speed of eating (Root and de Rochemont 1976) in the restaurants that would crop up to feed increasingly mobile Americans. National park eateries fueled a populace conversant with radical ideas from the likes of George Cheyne, John Wesley, John Harvey Kellogg, and J.I. Rodale associating spiritual and physical health with dietary choices—seekers whose time communing with nature in the parks required them to make distinctive choices about what they ate. Restaurant dining in national parks "made eating a form of entertainment and an object of conspicuous consumption as well as sustenance" (Lobel 2014, 6).

At the same time, parks quickly became unequipped to deal with the growing visitation rates facilitated by the opening of mass transit lines. Concessioners, eager to attract more well-to-do visitors, began opening grand, exclusive, luxury park lodges (Carr 1998) that would feature opportunities for opulent dining—at considerable cost, given the logistics of bringing fancy food to remote locations—in restaurants with priceless views. In Yellowstone, the Old Faithful Inn opened in 1904, and in the Grand Canyon, El Tovar in 1905, where visitors to the sizeable dining room "could dine on fresh salmon from the Pacific Coast, California fruit, Michigan celery, Camembert cheese, and other delectables of industrial America" (Rothman 1998, 59). Entrepreneur Fred Harvey, known unironically as "The Civilizer of the West," set up America's first restaurant chain, the Harvey Houses, alongside railway routes in the Southwest, many in national parks, including El Tovar. Harvey hired attractive young women as hostesses—"Harvey Girls"—whose competence and agreeableness were expected to have a civilizing influence on the rough, largely male clientele of the era (Fried 2010). Harvey's "Indian Department" promoted Indigenous arts and crafts and catered, through its "Indian Detours" program, to a tourist clientele hungry for cultural interaction with the so-called "forgotten people." Despite their interest in civilizing the public through enlightened dining and arts and culture tourism, concessioners often did not have public interests top of mind, and those running the parks cooperated with concessioners at the public's expense. For example,

> not until 1928 did the Interior Department solicitor emphatically rule that local beef being brought into the park for public consumption must be inspected, even though such a regulation would add ten cents a pound to the 35,000 pounds being imported each year.
>
> *(Bartlett 1985, 202)*

This concern for catering to those with more discerning tastes informed the motives of the men who established the NPS a decade later. Coney Island and Niagara Falls, sites of mass culture that provided working-class people with accessible recreation (Busà 2012), haunted their imagination as they set their vision for park concessions against these cautionary tales of the tawdry, the tacky, and the overrun. In an oft-told tale, borax magnate and publicity savant Stephen Mather got his job as first Director of the NPS in 1916, two years after writing to Secretary of the Interior Franklin K. Lane to complain about the terrible food and accommodations in Yosemite Valley; Lane told him to run the parks himself if he didn't like it.

A descendant of well-known Puritan minister and Harvard College president Increase Mather, Stephen Mather was widely admired. A good-looking live wire, full of energy and drive, described as "the popular dream of a man of distinction" (Shankland 1951, 8), Mather was a well-intended patrician. His immediate agenda upon his appointment as NPS Director was brimming: get Congress interested and willing to increase appropriations and to authorize a national park bureau, get said bureau up and running while increasing public interest by making travel easier, add more sites while keeping out undesirable ones, and get rid of private holdings, among other things (Shankland 1951, 56).

Although he is revered for his pioneering leadership of the NPS, Mather's view of the common people was arguably somewhat more instrumental than is commonly understood. Mather was credited with consciously shifting the focus of the parks from the elite to middle-class visitors, forging citizens who could provide political support for the parks out of everyday tourists (Sheail 2010). Presaging neoliberal themes decades before their widespread articulation, Mather was very interested in the ability of the parks to create better citizens: "He is a better citizen with a keener appreciation of the privilege of living here who has toured the national parks," he wrote in his 1920 Report as Director of the NPS (Mather 1920, 13–14).

When famed park lodge architect Gilbert Stanley Underwood expressed dismay at how those traveling by car (the "tin-canners") often left camps and picnic grounds strewn with litter, Mather replied, "What if they do? They own as much of the parks as anybody else. We can pick up the tin cans. It's a cheap way to make better citizens" (Shankland 1951, 161). Mather was convinced that time spent in the parks would make visitors better citizens, more thoroughly possessed of the land, and he didn't mind picking up their litter if it meant the public would bankroll the conservation of parklands. For greater numbers of tourists to be able to access the treasures of the national parks, the early NPS modeled the parks as resorts to keep the land from being mined, grazed, or logged (Mark 2005). Mather supported commercial development in the parks to accommodate more visitors; this "resort-friendly" stance empowered concessioners, whose amenities and recreational opportunities attracted tourists eager for engaging recreation opportunities in beautiful places.

However, not just any concessioner would do. One of Mather's chief goals was "to sweep the superfluous and tacky concessions from the parks" (Zaslowsky and Watkins 1994, 23). In 1923, Mather visited Coney Island with park landscape architect Gilbert Stanley Underwood, who would later design Yosemite's famed Ahwahnee Hotel:

> The two men spent the day wandering through the crowds, stuffing themselves with hot dogs and raw onions, and washing it all down with orange pop. At one point Mather turned to Underwood and said, 'This is exactly what we don't want in the national parks. Lots of people seem to like it and if they do, they ought to have it, but not in the national parks. Our job in the National Park Service is to keep the national parks as close to what God made them and as far as we can from a horror like this.'
>
> *(Zaslowsky and Watkins 1994, 26)*

These feelings about Coney Island were not unique to Mather but echoed by Secretary of the Interior Harold Ickes, who more than a decade later raised concern about the effects of the infrastructure and commercial development that had supported a surge in tourist visitation in the 1920s and 1930s:

> I have told them [Park Service officials] that we do not want any Coney Islands and that the parks are for those who will appreciate them and not merely for hordes of tourists who dash through them at break-neck speed in order to be able to say that they have been to Glacier or Yellowstone or some other park.
>
> *(Ickes 1938, cited in Mackintosh 1985, 82–83)*

This sense that concessions should be shaped to keep "lots of people" with vulgar tastes at bay—to offer what Urry (1995) has described as a social differentiation-enhancing "positional good"—speaks to the founding premise of the NPS.

Lest one think that Mather's ideas about the national parks as spaces of distinction are anomalous, the writings of second NPS Director Horace M. Albright are instructive. Albright and Taylor (1928) declared that national park visitors were of two types: "dudes" (also known as "couponers"; see Bartlett 1985), who came by train and "motor stage" and stayed at the hotels and lodges, and "sagebrushers," who came by covered wagon (or later automobile) and proceeded in a more do-it-yourself fashion, camping and cooking for themselves. The dude, initially a figure held in slight comic contempt by the more rough-and-tumble locals, was a well-dressed, clean tourist of means who flitted from concession to concession. Albright noted that by 1928, the term had become one of distinction, proudly owned by the tourists themselves (and preserved for posterity in "dude ranches," where wealthy tourists could go to have a try at authentic ranch living). Mather and Albright presided over an NPS with young, initially genteel luxury hotels providing tasteful accommodations to moneyed travelers, reflecting the early NPS commitment to making the parks attractive for use by citizens who would throw their support behind preserving the lands they had so enjoyed. Both Mather and Albright elevated "'park values,' reliability, and public welfare above profits made on government property," believing that "the park visitor deserved decent lodging and basic amenities, not hucksters clamoring to make a sale" (Keller and Turek 1998, 149). Hucksters and their easy marks were to remain at Niagara Falls and Coney Island.[2]

The Rise of Park Concessions

Mather, with his successful business background, did not like the state of concessions at the 14 existing national parks when he was appointed first director of the NPS in 1916. There was a somewhat chaotic array of hotels, camps, restaurants, and stagecoach and railroad lines in the parks, run by aggressive entrepreneurs. Mather preferred a more economical and orderly system that prevented the duplication of basic overhead among venues that took up too much space and provided low-quality, inconsistent service (Shankland 1951). He decided to move to a system where just one operator per park would be licensed to do all the concessions and to do so in a closely regulated way: a strictly regulated monopoly would protect both the concessionaire and the public, he reasoned (Shankland 1951).

This new model of concessions served a rapidly growing class of park tourists, whose mass market buying power came to outweigh the patronage of the elite (Blodgett 1990, 131). The shift to more democratic park access was enabled not only by the rise of the automobile but also by the rise of concessions. As more visitors started driving themselves, they didn't need the sightseeing tours and luxury hotels previously relied upon by wealthy, train-bound tourists. Starting in the late 1920s, using private funds invested in monopolistic franchises (Carr 2007, 230), concessioners obtained government approval and began to invest in facilities to cater to this less affluent but more populous crowd. Coffee shops, grills, and cafeterias cropped up alongside the formal dining rooms of the grand lodges, offering food at lower price points and greater speed, evidence of a new kind of cultural "massification" (Peterson 1997, 85). Finn notes that a new culinary ethos, idealizing simplicity, frugality, and reliability, emerged in the 1930s–1940s as middle-class Americans, emerging from the Great Depression into relative material comfort, embraced not the kinds of distinction prized by their Gilded Age forebears, but rather "all that was mainstream. Distinctive foods—whether they were gourmet, diet, natural, or foreign—were increasingly viewed with suspicion and disdain" (Finn 2017, 79). This dominant culinary ethos both shaped and is reflected in the foods and eating styles available within national parks at the time and has cast a long shadow on park concessions even today.

Things slowed down with concessions during World War II (WWII), when the NPS was accused of neglecting the parks. Immediately post-WWII, there was extremely strong regulation of concessioners after park visitations increased and strained concessioners couldn't keep up with demand. As a result, "between 1946 and 1950, twenty concession contracts expired with the Department of the Interior unable to find new concessioners" (Mantell 1979, 18n113). An agreement was made in 1950 between the concessioners and the government that safeguarded the investment of the concessioners while at the same time protecting the public from overly high prices and inadequate facilities (Mantell 1979, 20).

By the mid-1950s, tourists' complaints about the parks centered on over-visitation, inadequate lodging, and not enough restaurants (Carr 2007, 89), and visitors were most interested in remedying these shortfalls with affordable and convenient options. Conservation-minded parks supporters critiqued the way American tourists, still largely white but increasingly middle and even working class, with short attention spans and hungry for ever-new ways to recreate, were using the parks. "National Parks Association Member Paul Shepherd Jr. insisted that 'something was amiss' and felt that 'a larger majority of visitors [were] unaware of the peculiar meaning ... of the parks'" (Carr 2007, 54). Against this view of tourists as a pervasive threat—the ignorant, anti-Romantic visitor—came the controversial Mission 66 initiative, an effort proposed by NPS Director Conrad Wirth in 1955 to accommodate the increase in park visitors made possible by postwar prosperity through the development of infrastructure and the improvement of interpretive services. The goal was to "fix" the parks by 1966, the 50th anniversary of the founding of the NPS. This fix meant that the parks, through roads, restaurants, and motels, could cater more democratically to visitors who were nonetheless reasonably prosperous—and it could, through interpretive services at visitor centers (which are typically adjacent to cafeterias and grills, yet somehow a world away), teach visitors about the "peculiar meaning" of the parks that previous generations of far wealthier travelers knew so well.

Part of the "fix," as the work of Mission 66 was ending, was the development of a Concessions Policy Act (CPA) that was regarded with delight by concessioners. Approved by Congress in 1965 to guide the management of concessions, the CPA aimed to encourage development, use, and profitability of concession facilities (Mantell 1979, 28). Congress granted concessioners atypical rights, including terms of up to 30 years, preference in contract renewal, and a right of compensation for property improvements they made, to compensate for the "adverse commercial conditions" they needed to operate under in remote park spaces (Coggins and Glicksman 1997, 745). After Mission 66 expired, the NPS moved to individual park master planning, and since then, the concessioners have become a driving force in trying to expand and see their services used more (Mantell 1979, 30).

Park Concessions and the Return of the Cosmopolitan

With many of the 30-year contracts promised in the 1965 act having been tagged with brief extensions, and many set to expire in the mid- to late 1990s, controversies abounded over existing NPS concession policies. Complaints included "overdevelopment of visitor facilities, monopolistic concessionaire arrangements and practices, artificial stimulation of visitor demand, meager financial return to the government, inadequate facility maintenance, administrative secrecy in the contracting processes, and general lack of competition" (Coggins and Glicksman 1997, 729–30). Yet a majority of the public said they felt that restaurants belonged in the parks (Haas and Wakefield 1998).

These concerns led Congress to pass the National Parks Omnibus Management Act of 1998, part of which was meant to improve the visitor experience by stimulating competition. Restaurant concessions benefitted from lower franchise fees, as the food and beverage business was understood to be less profitable than lodging or transportation concessions (Martin 2005). Mom-and-pop concessioners—those grossing $500,000 or less—maintained their preferential contract renewal (Martin 2005). Although this protection extended to about three-fourths of the concession contracts in the parks, it did not cover too many restaurants, which tend to be run by large conglomerates since the middle of the twentieth century. According to the Government Accountability Office (GAO), "in 2015, 20 concessions contracts with the largest gross revenues accounted for about two thirds of the total revenues generated under all concessions contracts" (2017). GAO clarifies that the largest contracts, like those at Yosemite and Yellowstone, tend to offer lodging, food, and retail. Part of the problem is structural: smaller concessioners have a hard time competing for large contracts because they can't afford to purchase high leasehold surrender interest (LSI) balances from the outgoing concessioner, and although the government has bought down these balances in some cases and limited the amount of LSI (value of capital improvements made by a concessioner, adjusted for depreciation and inflation) that can be incurred to induce competition (Government Accountability Office 2017), it's still an uphill battle. Likewise, as concessioners must now prepare complex and labor-intensive prospectuses to bid for contracts, only a few big corporations can afford the time and expertise to assess whether the bid is worth the effort (Scott and Scott 2019).

By 2016, the NPS was managing 488 concession contracts (111 of them for food service), generating gross revenues of $1.4 billion annually and dealing with problems of lack of oversight of concessioner performance (Government Accountability Office 2017; Bryant 2019). At the ground level, hungry park visitors were greeted primarily by industrial food (think fast and convenient, from the back of a Sysco truck: pizza, burgers, hot dogs, cold sandwiches, french fries, and iceberg lettuce salads) in cafeterias, grills, camp stores, and table-service restaurants, served by highly profitable conglomerates. Park visitation is at record high levels, and The Cosmopolitan has returned—although this time, The Cosmopolitan is a type of visitor, not an establishment. Mass park tourism has been accompanied by an increase in efforts at social differentiation among visitors, and a particular stripe of them see food as a site of self-making and a source of distinction. As the traditional comfort foods typically offered at national park eateries are associated with lower socioeconomic status (Hupkens, Knibbe, and Drop 2000, 112), the upscale, health-conscious, modern park visitor is left with poor choices. Thus, they reflect the kind of anxiety over the industrial food situation in the parks that I overheard in Carlsbad Caverns and that I shared in the Introduction.

Today, as Americans leave vacation days unused each year, the national parks solve a moral problem for cosmopolitan visitors. They allow a certain class of visitors to claim redemptive value and moral virtue for their healthy leisure activities undertaken while actively communing with nature. Conveniently, the national parks also offer opportunities for reflection on the ways in which culture has damaged nature (Urry 1995, 226). For cosmopolitans, whose style of recreation otherwise meshes well with what the national parks offer—what Joseph Sax has described as "opportunity for the internalization of satisfaction, based on personal knowledge, individual style and expression, autonomy, and a setting rich enough in its complexity to elicit distinctive personal responses" (1980, 57)—the food experiences available in the parks are sorely lacking. Sax argued that the primary function of national parks should be to stimulate people's ability for "more challenging and demanding" reflective recreation and didn't mind that this was at odds with increasing park accessibility, concluding that "We need a willingness to value a certain kind of experience highly enough that we are prepared to have fewer opportunities for access in exchange for a different sort of experience when we do get access" (Sax 1980, 83).

But what are the implications for concessions? According to Sax,

> Supportive services—supply stores, unpretentious restaurants associated with hotels, and gas stations in more remote parks—are perfectly appropriate. What do not belong in such places are facilities that are attractions in themselves, lures that have nothing to do with facilitating an experience of the natural resources around which the area has been established.
>
> *(Sax 1980, 88)*

Sax takes Jackson Lake Lodge, in Grand Teton National Park, "with its fancy shops, swimming pool, and elegant restaurant" (88) as case in point.

Sax seems to miss how much more the elegant restaurants are aiding with the park's interpretive mission, facilitating an experience of local foods, than "unpretentious restaurants" of the typical fast-food order do, especially when these are run by multinational conglomerates; this is a theme I explore in more depth in Chapter 5. The cosmopolitans have a point when they insinuate that nothing in the Carlsbad Caverns cafeteria is going to stimulate reflection, allow for the expression of individual style and expression, or elicit distinctive personal responses. Too much accessibility threatens conservation as it leads to overrun parks, but lackluster industrial food in corporate cafeterias aimed at the "unpretentious" threatens craft consumption; the returned Cosmopolitan would say to Sax, ye of little faith, that he's missed the mark on his concessions-related musings. Sax encourages forms of park tourism that give the tourist a distinctive sense of the destination and serve the local population (1980, 107), yet he fails to see how elevated attention to national park foodways can accomplish these very ends to the delight of cosmopolitan visitors (and to the more questionable benefit of everyone else). How might we take advantage of cosmopolitan tourists' increasing dissatisfaction with standardized food tourism offerings that deny them the active engagement and learning, cultural capital, and social status (Laing and Frost 2015, 190) they so crave to push for something that better serves the land and the people most in need? The later chapters will provide some food for thought regarding these questions, but in the meantime, I will focus more deeply in Chapter 3 on the "local population" that so concerned Sax.

Notes

1 Although widely cited as the first national park in the world, Yellowstone (1872) may be the new kid on the block next to Bogd Khan Uul National Park in Mongolia, which dates to 1783.
2 This sentiment was not unique to the first half of the twentieth century. Garrett Hardin, opining in 1969 that demand is ruining the wilderness, echoes Stephen Mather's picking on Coney Island years earlier: "The carrying capacity of a Coney Island (for those who like it, and there are such people) is very high; the carrying capacity of wilderness … is very low" (Hardin 1969, 22). He clarifies that for Coney Island it might be 100 ppl/acre, and for wilderness, 1 person/sq mile. The assertion that "and there are such people," offered almost as a casual aside, bespeaks an elite preoccupation with the vulgar tastes of the masses that has plagued environmentalist opinion over the lifetime of the US national parks.

References

Albright, Horace M. and Frank J. Taylor. 1928. "Dudes and Sagebrushers." *Oh, Ranger! A Book About the National Parks*. Palo Alto, CA: Stanford University Press. https://www.nps.gov/parkhistory/online_books/albright3/chap2.htm.

Bartlett, Richard A. 1985. *Yellowstone: A Wilderness Besieged*. Tucson: University of Arizona Press.

Baylor University. "Native Americans Modified American Landscape Years Prior to Arrival of Europeans." *ScienceDaily*, March 22, 2011. www.sciencedaily.com/releases/2011/03/110321134617.htm.

Blodgett, Peter J. 1990. "Visiting 'The Realm of Wonder': Yosemite and the Business of Tourism, 1855–1916." *California History* 69 (2): 118–33.

Bryant, Kelsey. 2019. "Concessions Causing Detrimental Impacts on the Original Vision of National Parks." Student Note. *Kentucky Law Journal* 107. https://www.kentuckylawjournal.org/online-originals/index.php/2019/06/13/concessions-causing-detrimental-impacts-on-the-original-vision-of-national-parks.

Busà, Alessandro. 2012. "Rezoning Coney Island: A History of Decline and Revival, of Heroes and Villains at the 'People's Playground.'" In *The World in Brooklyn: Gentrification,*

Immigration and Ethnic Politics in a Global City, edited by Judith N. DeSena and Timothy Shortell, 147–84. Lanham, MD: Lexington.

Carr, Ethan. 1998. *Wilderness by Design: Landscape Architecture and the National Park Service*. Lincoln: University of Nebraska Press.

Carr, Ethan. 2007. *Mission 66: Modernism and the National Park Dilemma*. Amherst: University of Massachusetts Press.

Coggins, George Cameron and Robert L. Glicksman. 1997. "Concessions Law and Policy in the National Park System." *Denver Law Review* 74 (3). https://digitalcommons.du.edu/cgi/viewcontent.cgi?article=1988&context=dlr.

Cole, Thomas. 1836. "Essay on American Scenery." *American Monthly Magazine* 1: 1–12.

Cronon, William. 1995. "The Trouble with Wilderness." *The New York Times*, August 13. 6: 42. https://www.nytimes.com/1995/08/13/magazine/the-trouble-with-wilderness.html.

Finn, S. Margot. 2017. *Discriminating Taste: How Class Anxiety Created the American Food Revolution*. New Brunswick: Rutgers University Press.

Fried, Stephen. 2010. *Appetite for America: Fred Harvey and the Business of Civilizing the West, One Meal at a Time*. New York: Bantam.

Government Accountability Office. 2017. "National Park Services: Concessions Program Has Made Changes in Several Areas, But Challenges Remain." Report to Congressional Requesters. https://www.gao.gov/assets/gao-17-302.pdf.

Green, Linda W. 1987. *Yosemite: The Park and Its Resources--a History of the Discovery, Management, and Physical Development of Yosemite National Park, California* (Denver: National Park Service). Historic Resource Study series. Government Printing Office. http://www.yosemite.ca.us/library/yosemite_resources/state_grant.html.

Haas, Glenn E. and Timothy J. Wakefield. 1998. *National Parks and the American Public: A National Public Opinion Survey on the National Park System*. Washington, DC and Fort Collins: National Parks and Conservation Association and Colorado State University.

Haraway, Donna. 1984. "Teddy Bear Patriarchy: Taxidermy in the Garden of Eden, New York City, 1908–1936." *Social Text* 11: 20–64. https://doi.org/10.2307/466593.

Hardin, Garrett. 1969. "The Economics of Wilderness." *Natural History* 78 (6): 20–27.

Hupkens, Christianne L.H., Ronald A. Knibbe, and Maria J. Drop. 2000. "Social Class Differences in Food Consumption: The Explanatory Value of Permissiveness and Health and Cost Considerations." *European Journal of Public Health* 10 (2): 108–13.

Janiskee, Bob. 2011. "National Park History: Yosemite's Cosmopolitan Bathhouse and Saloon." *National Parks Traveler*, September 20. https://www.nationalparktraveler.org/2011/09/national-park-history-yosemites-cosmopolitan-bathhouse-saloon-1871-18848780.

Keller, Robert H. and Michael F. Turek. 1998. *American Indians and National Parks*. Tucson: University of Arizona Press.

Kimmel, Michael. 2008. *Manhood in America: A Cultural History*, 2nd ed. Oxford: Oxford University Press.

Klein, Lauren F. 2020. *An Archive of Taste: Race and Eating in the Early United States*. Minneapolis: University of Minnesota Press.

Laing, Jennifer and Warwick Frost. 2015. "The New Food Explorer: Beyond the Experience Economy." In *The Future of Food Tourism: Foodies, Experiences, Exclusivity, Visions and Political Capital*, edited by Ian Yeoman, Una McMahon-Beattie, Kevin Fields, Julia N. Albrecht, and Kevin Meethan, 177–93. Bristol: Channel View.

Lamont, Michèle. 1992. *Money, Morals and Manners: The Culture of the French and American Upper-Middle Class*. Chicago, IL: University of Chicago Press.

Lobel, Cindy R. 2014. *Urban Appetites: Food and Culture in 19th Century New York.* Chicago, IL: University of Chicago Press.

Mackintosh, Barry. 1985. "Harold L. Ickes and the National Park Service." *Journal of Forest History* 29 (2): 78–84. http://npshistory.com/publications/jfh-v29n2-78-84.pdf.

Mantell, Michael. 1979. "Preservations and Use: Concessions in the National Parks." *Ecology Law Quarterly* 8 (1): 1–54.

Mark, Stephen R. 2005. *Preserving the Living Past: John C. Merriam's Legacy in the State and National Parks.* Berkeley: University of California Press.

Martin, Stephen P. 2005. "Statement of Stephen P. Martin, Deputy Director, National Park Service, Department of the Interior, before the Subcommittee on National Parks, House Committee on Resources, regarding the implementation of the National Park Service Concessions Management Improvement Act of 1998." April 6, 2005. https://www.doi.gov/ocl/nps-concessions-act.

Mather, Stephen. 1920. "Report of the Director of the National Park Service to the Secretary of the Interior for the Fiscal Year Ended June 30, 1920 and the Travel Season 1920, 13–14." Washington, DC: Government Printing Office. http://npshistory.com/publications/annual_reports/director/1920.pdf.

National Park Service. 2021. "Concessions History." Yosemite National Park, last updated March 5, 2021. https://www.nps.gov/yose/learn/historyculture/concessions-history.htm.

National Parks Omnibus Management Act of 1998. 1998. Public Law 105-391, November 13, 1998. 112 STAT. 3497. https://www.congress.gov/105/plaws/publ391/PLAW-105publ391.pdf.

"No Longer Sodom-by-the-Sea: Coney Island Made Respectable by Brooklyn Police." *The New York Times*, May 7, 1894. https://timesmachine.nytimes.com/timesmachine/1894/05/07/106903632.html?pageNumber=1.

Peterson, Richard A. 1997. "The Rise and Fall of Highbrow Food Snobbery as a Status Marker." *Poetics: Journal of Empirical Research on Culture, the Media and the Arts* 25 (2): 75–92.

Roos, Dave. 2020. "Native Americans Used Fire to Protect and Cultivate Land." *History.com*, September 18, last updated July 30, 2021. https://www.history.com/news/native-american-wildfires.

Roosevelt, Edith Kermit. 1957. "Coney Isle Fishing for Way to Regain Its Lost Glamour." *Buffalo Evening News*, June 1. https://fultonhistory.com/Newspaper%2024/Buffalo%20NY%20Evening%20News/Buffalo%20NY%20Evening%20News%201957/Buffalo%20NY%20Evening%20News%201957%20-%203502.pdf.

Root, Waverly and Richard de Rochemont. 1976. *Eating in America: A History.* New York: Ecco Press.

Rothman, Hal K. 1998. *Devil's Bargains: Tourism in the 20th Century American West.* Lawrence: University of Kansas Press.

Sax, Joseph L. 1980. *Mountains Without Handrails: Reflections on the National Parks.* Ann Arbor: University of Michigan Press.

Schmitt, Peter J. 1969. *Back to Nature: The Arcadian Myth in Urban America.* London: Oxford University Press.

Scott, David and Kay Scott. 2019. "Consolidation in Managing the National Park Lodges." *NationalParksTraveler.org*, March 15. https://www.nationalparkstraveler.org/2019/03/consolidation-managing-national-park-lodges.

Shankland, Robert. 1951. *Steve Mather of the National Parks.* New York: Knopf.

Sheail, John. 2010. *Nature's Spectacle: The World's First National Parks and Protected Places*. London: Earthscan.

Solnit, Rebecca. 2000. *Savage Dreams: A Journey into the Hidden Wars of the American West*. Berkeley: University of California Press.

"The Attempt to Save Niagara." 1885. *The New York Times*, April 5: 11. https://timesmachine.nytimes.com/timesmachine/1885/04/05/106173631.html?pageNumber=11.

Urry, John. 1995. *Consuming Places*. London: Routledge.

Wallach, Jennifer Jensen. 2013. *How America Eats: A Social History of U.S Food and Culture*. Lanham, MD: Rowman and Littlefield.

Williams, G.W. 2000. "Introduction to Aboriginal Fire Use in North America." *Fire Management Today* 60 (3): 8–12.

Williams, G.W. 2002. "Aboriginal Use of Fire: Are There Any 'Natural' Communities?" In *Wilderness and Political Ecology: Aboriginal Influences and the Original State of Nature*, edited by C.E. Kay and R.T. Simmons, 179–214. Salt Lake City: University of Utah Press.

Zaslowsky, Dyan and T.H. Watkins. 1994. *These American Lands: Parks, Wilderness, and the Public Lands*. Washington, DC: Island Press.

3

INDIGENEITY AND EATING IN US NATIONAL PARKS

US national parks are known for spectacular vistas, and for wholesome recreation opportunities targeting contemporary middle-class (white) tourists, who are today encouraged to "find your park." As argued in Chapter 2, cosmopolitanism serves as a lens that helps us to understand, contextualize, and evaluate many elements of the experience of eating in the US national parks. However, the enabling conditions of cosmopolitanism are obscured in most discussions of park foodways. This chapter attempts to illuminate how the Romantic consumption discussed in the Introduction occurs on the backs of displaced others, whose own forms of food production and consumption were, and continue to be, wildly disrupted and fundamentally transformed, manifesting a settler colonialist structure (Wolfe 2006). As such, this chapter explores the cultural history and the current situation of Indigenous foodways in today's parklands.[1]

Worldwide, more than half of all national parks and protected areas occupy unceded Indigenous lands (Zeppel 2009, 259). It is little surprise that the lands of Indigenous people, whose rights to self-determination have been historically quashed by settler colonialism, have been prime sites for parks in the US, where the history of national parks aligns almost to the year with the US government's divestiture of Indigenous people from their sovereign status: American Indian treaty making ended in 1871, just before the establishment of Yellowstone by Congress in 1872.

Parks remain sites of struggle for American Indian food sovereignty. In many cases, what made the lands worth fighting for by Indigenous peoples was not only their familiarity, their sacredness, or their beauty but also their status as productive foodscapes. Prior to settler contact, for American Indians west of the Mississippi River—in the parts of the country with the densest concentration of what is now US national parkland—the chief food use of these lands was hunting, fishing, and gathering, in line with various tribes' awareness of local ecologies (Hufford 1986).

DOI: 10.4324/9781003455516-4

Some bands, like the Tukudika Shoshone (Sheep Eater), lived in what would later become Yellowstone National Park year-round, while many other groups of American Indians, including the Bannock, Blackfeet, Crow, and Nez Perce, used natural resources in today's parklands for hunting, gathering, and trading food during seasonal migration. They also used the land to make tools and weapons, and for spiritual ceremonies (Patterson 2022). These land uses in Yellowstone are representative of Indigenous people's land uses in the West, and they manifest what Clayton (2020) has described as the connection between the individual and the wider natural world. Although today, many tourists imagine places like Yellowstone as uninhabited, this vision obscures the reality that today's parkland was used by Indigenous people and their predecessors for 11,000 years (Zeppel 2009, 263).

Wilderness and Restoration

The people of the US have long operated under delusions about wilderness that cloak the reality of these lands as historical foodscapes. European settlers first industrialized the eastern US in the early nineteenth century, rapidly building crowded, dirty cities from which they needed an escape. Expelled from the Garden of Eden, settlers in some ways understood their cheek-by-jowl lives in hostile cities as punishment for original sin (Meeker 1973, 3). They craved experiences of nature that made them feel sanctified, before returning to their polluted city lives. Frederick Law Olmsted, father of American landscape architecture and noted conservationist and social critic, popularized the notion of recreation as vital for the economy, in that it helped workers to be more productive. Olmsted argued for helping people "unbend" by giving them something different—nature—to look at and ponder, as opposed to industrialized spaces, convinced that such respites would contribute to the nation's wealth—a seemingly necessary coping mechanism for the excesses of the Gilded Age. Olmsted emphasized the salutary psychological effects of this aesthetic experience and warned of the dire consequences of a nature deficit.

> The want of such occasional recreation often results in a class of disorders the characteristic quality of which is mental disability, sometimes taking the severe forms of softening of the brain, paralysis, palsy, monomania, or insanity, but more frequently of mental and nervous excitability, moroseness, melancholy or irascibility, incapacitating the subject for the proper exercise of the intellectual and moral forces.
>
> *(Olmsted 1865, np)*

The West was viewed as the ideal site for these restorative efforts. In fact, all but one of the first 15 areas designated as US national parks were in the West—and the one site chosen east of the Mississippi during that time period did not fare well, as Mackinac National Park in Michigan retained its designation for only 20 years.[2]

Not until nearly 50 years after the designation of Yellowstone as a national park in 1872 was another park east of the Mississippi, Acadia (then called Lafayette), designated. The striking east-west divide was attributable in part to population density, as more sparsely populated lands presented fewer political challenges for conservation. But it was not just population but also ideology, as the West was seen as a particularly tantalizing site for American self-making.

Western park visitors were not just escaping the busy industrial cities of the east coast and their workaday woes; after the Civil War, they were visiting a reinvented US, untarnished as the West was by the slavery, sectionalism, and states' rights battles that had marred the South and the East. Per Hal K. Rothman, "The revised national creation myth gave the West primacy in American life and thinking that grew from innocence and potential for reinvention, a prestige further marking the region's importance in a postindustrial world increasingly dependent on tourism" (1998, 14–15). The mythos of the West as a place to heal and revitalize from trials both personal and national has since fueled generations of national park tourism, but it depended—and still depends—on political subjugation. Olmsted argued that

> the power of scenery to affect men is, in a large way, proportionate to the degree of their civilization and the degree in which their taste has been cultivated. Among a thousand savages there will be a much smaller number who will show the least sign of being so affected than among a thousand persons taken from a civilized community.
>
> *(1865, np)*

Like many of his colleagues who argued in support of the establishment of national parks, Olmsted believed that so-called "savages" were too uncivilized to appreciate the lands on which they had lived for thousands of years—that they were incapable of wonder, the very ideology that offered Olmsted's ilk a way to imagine the lands of the West as an untouched "wilderness" devoid of human presence.

Although they were not European—and indeed they took great pains to set themselves apart from Europeans—bourgeois American national park founders and supporters employed representational strategies that allowed them to assert European hegemony, with conservation motives providing a cover of innocence. Described by Pratt as the "seeing-man," the white male protagonist of European landscape discourse was "he whose imperial eyes passively look out and possess" (Pratt 1992, 9). The national park founders were invested in a version of environmentalism that involved what Richard Grusin has described as "deployment of the ideology of nature's intrinsic value to further the social, cultural, or political interests of a dominant race, class, gender, or institutional formation" (2004, 2). Inspired by wonder and fueled by curiosity, they leveraged conservation on the backs of Indigenous peoples.

Preservation, Conservation, and Tourism in "Worthless" Lands

As settlers became very interested in conservation, others were interested in the profits that park tourism could bring, particularly to railroads and other concessioners (Bartlett 1985).

> Without hotels, roads, boats, trains, and restaurants, few people would be tempted to make the arduous journey to the parks, much less endure them for an extended stay. And without tourists in Yellowstone and Glacier, the Park Service would have a difficult time justifying its existence to a Congress suspicious of the long-term benefits of land preservation.
>
> *(Burnham 2013, 78)*

Tourists and their dollars were considered essential to sustaining these sites.

Early park advocacy, driven by coexisting desires for conservation and tourist profits, required the advocates to engage with the lands in question as Indigenous foodscapes. For instance, Frederick Billings, president of Northern Pacific Railroad, played a significant role in Yellowstone's founding and development as a national park. He was attracted because it would require building new railroad lines and would open new markets. Mark Spence notes that

> Billings was an aggressive advocate for the further reduction of tribal landholdings. He even boasted that railroads could restrict Indians to their diminished reservations because the prairie fires caused by trains proved an effective, if somewhat accidental, measure for driving away the game that attracted native hunters to their recently ceded lands.
>
> *(Spence 1999, 37)*

In cases like this, we start to clearly see the dawning recognition of national parklands as Indigenous foodscapes.

John Muir, the influential naturalist known as the "Father of National Parks" for his substantial wilderness preservation efforts, saw tourism as a preferable alternative to grazing or commercial clear-cutting. Muir was particularly troubled by what might today be jokingly described as "Big Sheep"—grazing interests that exploited the land as their ruminants overgrazed, degrading vegetation and soils, with lasting negative ecological consequences (Muir 1890, np). Tourists, in contrast, were tolerable, so Muir wrote popular pieces for them, teaching them how to see the natural world that he had long adored (Mark and Hall 2009, 90). In what may be understood as a sort of "the devil you know" maneuver, Muir aligned with Southern Pacific Railroad magnate William H. Harriman in lobbying for federal control over Yosemite parkland, calculating that whatever ecological damage the railroads and the tourists they ferried might bring, they were offset by the benefits of containing the destruction caused by an overpopulation of poorly regulated concessioners and other commercial interests (Mark and Hall 2009, 98). In Muir's mind, the presence of increasing numbers of tourists would ensure ongoing preservation.

To win both popular and governmental approval for the development of national parks in a burgeoning economy concerned with growth, settler advocates had to prove (or manufacture) the idea that the lands the Indigenous peoples occupied and used were "worthless," an idea that could justify paternalistically moving Indigenous peoples off *and* conserving the lands. Surprisingly, conservation and preservation victories hinged on what lands lacked, rather than what features they had. Lands absent abundant material resources of minerals, water, timber, or appropriate terrain for grazing or agriculture were prime candidates for conservation. Preservation and tourism were enabled only when lands were declared worthless as production spaces for settler food—when, for instance, the soil would not facilitate cultivation, nor would the extreme weather enable ranching. An 1872 report to Congress by the Committee on Public Lands supported the withdrawal of the land now known as "Yellowstone National Park" from settlement, occupancy, or sale, noting that "the entire area…is not susceptible of cultivation with any degree of certainty, and the winters would be too severe for stock-raising" (Dunnell 1872). The report concluded that setting aside the Yellowstone land would "take nothing from the value of the public domain, and is no pecuniary loss to the Government, but will be regarded by the entire civilized world as a step of progress and an honor to Congress and the nation" (Dunnell 1872). Rescued from their alleged uselessness, the nonfood-producing "worthless" lands, now preserved as parks, were made productive through railroad-linked tourism opportunities (Hall and Frost 2009, 59).

Settler park advocates perceived of Indigenous peoples' presence as a danger to the lands—their fires and hunting, fishing, and harvesting practices were understood (incorrectly) as contrary to genuine conservation. American Indians were also seen as a danger to white tourists: violent conflicts between early park visitors and Indigenous peoples (sometimes themselves fleeing armed settler brigades) were bad publicity for the tourist enterprise. Although Richard Grove notes that "indigenous strategies for environmental management on a small scale, often involving a considerable understanding of environmental processes, had existed in many parts of the world since time immemorial" (Grove 1995, 6), later settler advocates misread Indigenous environmental management strategies as existential threats.

Extinguishing Indigenous Foodways

Indeed, within the first ten years of Yellowstone National Park's existence as such, complaints from white settlers and game hunters about Indians setting parts of the forest on fire and killing too many animals were prevalent.

To most champions of wilderness preservation, the best solution for protecting these areas was an old solution: the use of military force to keep native peoples on their reservations. Such a program would not only preserve wilderness but also fit nicely into ongoing efforts to "civilize" Indians by training them to become self-sufficient agriculturists.

(Spence 1999, 62)

Park managers doubled down on their impulses to protect game and prevent fires, even though it was white hunters who were killing animals for their hides only, unlike American Indian hunters, who killed for the meat (Spence 1999, 65).

The desire to protect game had its roots in proposals by early park advocates including John Muir and George Bird Grinnell, who were concerned that hunting by both whites and American Indians would deplete game stocks, endangering wildlife and robbing the land of its special character. Initially, Indigenous peoples were allowed to hunt inside the parks because they had previously established usufructuary rights—rights to use the lands they ceded for activities like hunting, gathering, and fishing—through treaties signed with settler societies, usually under dubious circumstances. For instance, in treaty council meetings in 1867 and 1868, Crow, Kiowa, Comanche, Cheyenne, and Arapaho leaders demanded and were granted usufructuary rights to off-reservation lands, and the US government in turn recognized each tribe's right to hunt game on these unoccupied lands, some of which later became national parklands (Spence 1999, 32). Although the establishment of the first national park marked a major act of Indigenous dispossession, the process had already begun in the years leading up to it. Even before these changes in food provisioning could play out, Indigenous people's treaty rights were eclipsed by new state laws forbidding hunting on public lands. Stephen T. Mather, founder of the National Park Service (NPS), regarded treaty rights to hunting as an existential threat to the federal government's exclusive jurisdiction, and hunting by Indigenous peoples became, in a word, the white whale of the NPS.

Alongside hunting, Indigenous fishing and harvesting on parklands have also been hot-button issues for park advocates (see Ruppert 2003; Kantor 2007; Lewis 2012; Wolfley 2016). Although many NPS units (national forests, lakeshores, seashores, and recreation areas) allow hunting, fishing, and harvesting (subject to state regulations), in the national parks themselves, these pursuits have historically been regarded as anti-preservationist and thus prohibited.[3] However, Indigenous cultures have understood their food procurement activities "not only as a form of natural regulation, but also as an expression of connection between an individual and the wider natural world" (Clayton 2020); deprived of the rations and tools promised in their treaty obligations, many Indigenous people have seen hunting, fishing, and harvesting on parklands as the obvious alternative to starvation or reliance on low-quality commodity foods.

Several cases pertaining to Indigenous food procurement in parkland have been argued before the US Supreme Court. *Ward v. Race Horse* (1896) sealed the hunting disenfranchisement of Bannock Tribe members in the same shameful week that it decided the historic *Plessy v. Ferguson* case. *Minnesota v. Mille Lacs Band of Chippewa Indians* (1999) and *Herrera v. Wyoming* (2019), more recent cases, have upheld usufructuary rights to Indigenous subsistence activities on parklands. It has taken well more than a century for the US government to restore some of the food procurement treaty rights granted in the 1860s. Recent measures like the 2016 NPS rule permitting application by American Indian tribes to harvest plants on parkland hang the tribes up in so much red tape that they continue to interfere with Indigenous people's right to subsist.

Even in parklands where bans on fishing, hunting, or gathering were enacted, they were often inequitably enforced, to the benefit of white tourists and concessioners, and at the expense of Indigenous peoples. For instance, the overfishing that supplied the whitefish at Glacier National Park's Great Northern lodges in the 1930s was done in contravention of park legislation that banned net fishing; there doesn't seem to be any American Indian involvement in the actual netting, which was done by hotel employees (Keller and Turek 1998, 59). Blackfeet Natives were not allowed to net fish or hunt for subsistence within the park, but park officials seem to have looked the other way when faced with the continued commercial fishing for Louis Hill's restaurants that served tourists hungry for a taste of the local.

Park advocates imagined American Indians as threats not just to the land, through their hunting, fishing, harvesting, and fires, but also to the physical safety of white visitors. Often, Indigenous people were understood as a menace by park managers and those standing to profit from park tourism. Blodgett describes the Mariposa Battalion, the militia that tracked and evicted Ahwahnechee and Chowchilla people in 1851, as "Yosemite's first white 'tourists,'" who "for the most part proved quite impervious to the valley's marvels" (Blodgett 1990, 119). They cleared the way for safe visits by landscape-gazers, whose guidebooks and diaries "reveal an almost complete lack of interest in Yellowstone's native history and little or no concern about Indian attack" (Spence 1999, 60). There were also several violent encounters between tourists and Nez Perce fleeing an army force in Yellowstone National Park in 1877 (Akerman 2020, 186). A recurring front-page section in the *New York Times* of the 1870s, "The Hostile Savages," ensured that the nation's educated citizens were well informed about the dangers posed by American Indians, who were described as violent torturers intent on humiliating their victims (September 10, 1872). Wrapped up in the desire to eradicate hunting and burning practices, then, was also a desire to keep Indigenous peoples out of the parks entirely, to ensure the safety of fellow settler travelers.

Frequently, efforts to banish Indigenous peoples from parklands hinged not on issues of safety but rather on taste or decorum. Lest anyone think that park conservation initiatives were for the good of all, consider what the Department of the Interior forest manager W. P. Hermann wrote to the Indian Office in 1898:

> The Grand Canyon of the Colorado River is becoming so renowned for its wonderful natural gorge scenery ... that it should be preserved for the everlasting

pleasure and instruction of our intelligent citizens as well as those of foreign countries. Henceforth, I deem it just and necessary to keep the wild and unappreciable Indian from off the Reserve.

(quoted in Keller and Turek 1998, 158)

American Indians were often billed as attractions for white park visitors who didn't mind a little simulacra mixed in with their "authentic" interactions with Tribal members. But outside of their participation in highly choreographed dances and tightly scripted tipi showings, likely motivated by their need for survival money in the new economy, Indigenous peoples in national parks were, at times, "repulsive and jarring" to well-off tourists, and thus were prime candidates for removal by the NPS (Keller and Turek 1998, 233). Even the Great Northern Railroad after World War II started to question the effectiveness of American Indians as a draw for tourists visiting Glacier National Park, as older Blackfeet who worked in this capacity were "rapidly dying off, and the younger generation are too modern in their habits…are hard to handle, and would not be suitable for entertainment purposes" (J.M. Budd and C.L. Finley, General Managers File, Great Northern Papers, cited in Keller and Turek 1998, 62).

Aversion to Indigenous people sometimes centered precisely on their foodways, as when environmentalist John Muir reflected on the Indians whom he found unclean and intermittently indolent during a visit to Yosemite in 1869. Muir wrote admiringly only of their "pure air and pure water," condemning "the grossness of their lives. Their food is mostly good berries, pine nuts, clover, lily bulbs, wild sheep, antelope, deer, grouse, sage hens, and the larvae of ants, wasps, bees, and other insects" (Muir 1911, 206). In short, exhibiting behavior deemed uncontrollable, foodways thought disgusting, and demonstrating an increasing unwillingness to perform their cultural identities in ways circumscribed by park operators and concessioners, Indigenous people became increasingly unwelcome in the parks established on land that had already been taken from them. When the way Indigenous people speak, act, and eat is taken as proof that they are uncivilized, it becomes ideologically easier to justify taking their land and allows this sleight of hand to be reimagined as for their own benefit. The legacy of this history manifests in anxious present-day white settler discourses about Indigenous peoples' foods as "unhealthy"—fried and fatty, boozy and bad.

Indigenous Hunger

Throughout the nineteenth century, American Indians were moved to reservations or restricted to a small portion of the land they could previously access, where settlers made efforts to "civilize" them via agriculture. The government, motivated by a belief that Indigenous peoples who became farmers would be able to better assimilate into white society, crafted policies to encourage agricultural work (Hurt

1987). The forced transition to agriculture by the Havasupai people illustrates the double bind for Indigenous peoples. As Philip Burnham notes, a lot of the traditional Havasupai hunting grounds were federally "protected" in 1893 as the Grand Canyon Forest Reserve. Deer hunting was forbidden for five years, which left the Havasupai peoples needing to recultivate the fields of recently deceased tribal members to have enough to eat (Burnham 2013, 75). We see in this example that food provisioning via agriculture, which we take for granted today, constituted an incursion against Indigenous food sovereignty when Indigenous peoples were left no other choice by virtue of the US government's park regulation.

In many cases, national parks are built on Indigenous peoples' hunger and food dependency. Thomas Jefferson endorsed putting American Indians on the dole—making them so desperate for food that they had no choice but to cede land from which they would otherwise not separate. This technique enabled the acquisition of the Indigenous peoples' lands that would be transformed into national parks. By way of example, consider the Piegan nation in northwestern Montana, starved by the 1880s by the near extinction of buffalo, infertile treaty land, and unsuccessful efforts at cattle herding, as well as by extremely stingy Congressional allotments amounting to "less than one ounce of beef per day for each agency Indian" (Burnham 2013, 40). Hundreds died of starvation before the Tribe yielded its land to the government, and its people were finally supplied with adequate food (Burnham 2013, 41). Likewise, Crow nation members in northeastern Yellowstone struggled with diminished game stocks and the destruction of their key food-gathering sites by miners. By the early 1880s, they too were almost completely dependent on agency rations (Spence 1999, 52), before ceding part of their land and then moving away.

An agricultural agenda was pressed into service as the most palatable solution to these problems. This agenda stems, in part, from nostalgia for the benevolent garden of Judeo-Christian origins, updated for the dawn of the nineteenth century and tied to the Jeffersonian legacy of romanticizing people's virtuous attachment to the land, which he believed gave them moral strength. Jefferson had in mind not the actual toilers but rather "farmer-landowners like himself, the pastoral gentlemen who owned and managed the American garden. Jefferson connected both moral virtue and political rights to land stewardship exercised by landowners" (Meeker 1973, 4). Against these settler visions, the nomadic way of life of most Indigenous peoples registered as entirely uncivilized to European immigrants with agrarian roots in the industrializing US. It was only agrarian life—settled, predictable, and tied to a specific plot of land—that was viewed as civilized (Burnham 2013, 17).

The insistence on agriculture in western reservation lands ill-suited to it destroyed the food traditions of many Indigenous peoples, to whom farming was culturally foreign anyway[4]:

Scorched-earth battle tactics utilized against Native people in the eighteenth and nineteenth centuries destroyed food supplies and the land from which they came in order to force Native people to become reliant on the American government. [...] Federal policies encouraged Native people on many reservations to farm on marginal lands, despite their histories of successful fishing and gathering practices. While some tribal communities traditionally practiced farming, others did not (such as Plains tribes and other communities across North America). The US and Canadian governments introduced farming projects to the latter in order to assimilate them, disrupt their hunting cultures, and expand the agricultural frontier—even as the best farmland was often usurped by non-Indians.

(Hoover and Mihesuah 2019, 5)

Agricultural efforts floundered, leaving American Indians dependent on the US government for the food rations that had been established as a stop-gap measure until the Indigenous people became self-sufficient farmers (US Department of State 1868; see also Rice 2022 for a more complex assessment of agricultural imperialism and Indian reservations). Not coincidentally, intensive agriculture has been cited as the most consequential invention in human history, with the fixed-harnessed plow cast as the technology with the most profoundly negative ecological impact (White 1967).[5] Although agriculture has some pretty stiff competition for this mantle—writing, the telegraph, automobiles, and microchips are no slouches either—the point is well taken: agriculture shifted the relationship of humans and the natural world from one of coexistence to one of mastery and exploitation—a shift particularly poorly suited to American Indian tribes who regarded nature with humility, reverence, and respect, and imagined themselves as deeply connected to the natural world.

A number of health and social problems have emerged in American Indian communities as a result of the disruption of traditional food systems. But health is not the only issue: as the availability of foods declined, so too have the stories, languages, cultural practices, interpersonal relationships, and outdoor activities implicated in those food systems. A tribal community's capacity for "collective continuance" and "comprehensive aims at robust living'" are hindered when the relationships that are part of traditional food cultures and economies are disrupted.

(Hoover and Mihesuah 2019, 7)

Recent food sovereignty efforts seek to remediate the harms caused when Indigenous peoples were separated from their traditional foodways (sometimes including less intensive forms of agriculture) by European American settlers. Food sovereignty goes beyond just agriculture: it means, following the 2007 Declaration of Nyéléni, "the right of peoples to healthy and culturally appropriate food produced through ecologically sound and sustainable methods, and their right to define their own food and agriculture systems." It involves reviving food-centered economic and cultural life and is a tool for economic opportunity and self-sufficiency in poor communities (Mertens 2021).

National park conservationists are likely to have something in common with Indigenous peoples, namely a belief that "capitalism's treadmill of production and assumptions of endless resource extraction are incompatible with nature, harmony, and balance" (Lindholm 2019, 160). The concession system in the parks, explored more deeply in other chapters, seems incompatible with this perspective, in that it depends on corporate profit and resource extraction and is poorly aligned with nature. In Chapter 6, I explore different food futures for the US national parks that prioritize environmental conservation and the livelihoods of Indigenous peoples. Is a food future in the parks possible that balances regulation around health and safety with food sovereignty? How would issues of labor, scale, volume, and palate affect operational viability? What would Indigenous foodways presented in national parks look like in highly regulated spaces, and who is the presumed audience in this equation? What is at stake in advocating a system that involves Indigenous people presenting, selling, or monetizing their cultural traditions? How is cosmopolitanism, the means of self-making for middle-class white travelers explored in the Introduction, implicated in the demand for access to Indigenous people's foodways in national parks? Is there a productive way forward?

Recentering Indigenous Foodways in the Parks

Eager readers may look for a conclusion to this chapter that documents the return to a thriving state of Indigenous foodways within the national park system; alas, that future remains to be written. Vibrant Indigenous foodways and commitments to food sovereignty are not merely historical issues, but they are yet to be recentered in the context of US national parks. The political stage is set for a recentering: as of this writing, the government leaders of both the Department of the Interior and the NPS are enrolled Tribal members who have expressed a commitment to co-management of the parks. Deb Haaland, US Secretary of the Interior, is the first Native American (Pueblo of Laguna) to serve as a Cabinet secretary. Haaland's past as a young single mother who started a small food business, Pueblo Salsa, while experiencing food insecurity and relying on food stamps to get by, is well documented (Hennessy 2021) and suggests that she has the life experience to imagine the possibilities of recentering Indigenous foodways in public lands. The NPS is led by Director Chuck Sams, who is Cayuse and Walla Walla and who is enrolled with the Confederated Tribes of the Umatilla Indian Reservation. In his first year, Sams has scored big with restorative justice efforts that involve repatriation of Tribal cultural items and the incorporation of narratives of affiliated Tribes into park visitor centers (Pennington 2022). Together, Sams and Haaland have championed the Tribal Homelands Initiative, a collaborative effort to improve federal stewardship of public lands, waters, and wildlife by incorporating Tribal capacity, expertise, and Indigenous knowledge into federal land and resources management (US Department of the Interior 2021).

On the Tribal side, prominent efforts at recentering foodways have been focused on the restoration of usufruct rights, less so on concession management or the marketing of Native American foods in park restaurants. Federally recognized Tribal governments that have successfully lobbied the NPS for agreements for plant collection include the Tohono O'odham Nation in Saguaro National Park (2018), the Eastern Band of Cherokee Indians in Great Smoky Mountains National Park (2019), and the Cherokee Nation in Buffalo National River (2019/2022). A fourth agreement will be finalized soon between the Pokagon Band of Potawatomi and Indiana Dunes National Park (National Park Service 2023). Other park-specific Tribal co-management agreements have restored food harvesting rights. For instance, Congress passed legislation in 2014 to authorize authorizing gull egg harvest by the Huna Tlingit in their traditional homeland of Glacier Bay National Park (Sams 2022). Likewise, the Nisqually Tribe is currently collaborating with Mount Rainier National Park to offer recommendations for administering traditional plant gathering in ways that minimizes ecological impact (Sams 2022).

Where park-specific enabling legislation provides Tribes the right of first refusal to provide visitor services, as between the Seminole Tribe of Florida and the Miccosukee Tribe of Indians of Florida in Big Cypress National Preserve, there has been no expression of interest in pursuing co-management agreements and there are currently none in place (Sams 2022). This reticence to provide direct visitor services within the parks may have good reason. To respond to the place-based market demands of cosmopolitan foodies, food producers "need to have a good command of the language, the aesthetic characteristics, and the discursive categories" these consumers observe (Parasecoli 2017, np). Yet when local Indigenous food producers lack the cultural capital and vocabulary needed to position themselves for the cosmopolitan market, experts and cultural intermediaries step in to validate their work and products and make them legible to cosmopolitan consumers (Parasecoli 2017, np).

The Cedar Pass Lodge and Dining Room in Badlands National Park provides an interesting case study of one of very few official park concessions that have been run by Indigenous people. In 1964, the NPS acquired a lodge and cabins (including a restaurant, gift shop, and gas station) that had operated privately in Badlands since before its designation as a unit of the NPS and, in turn, leased them to the Oglala Sioux Tribe. As many as 30,000 members of this tribe live on the Pine Ridge Indian Reservation, best known as the location of the Wounded Knee massacre, which occupies 2 million acres south of the park. The reservation comprises counties that are among the poorest in the US, and estimates put the unemployment rate at 80%–90% due to the lack of local economic development (John Milner Associates 2005, Section 3, 2). Ostensibly a boon for economic development, the Oglala Sioux Tribe operated the concessions for nearly 40 years, until they lost the contract in 2002 amidst financial woes (Scott and Scott 2019). The concession was taken over by Forever Resorts, which was subsequently acquired by Aramark;

travel guides still laud the restaurant there for its Sioux Indian tacos made with homemade frybread and frybread topped with wojapi, a traditional Lakota berry jam, regardless of who now profits from these sales.

Although the end of the proprietorship of Cedar Pass Lodge by Indigenous people could be interpreted as a cautionary tale about investing in industrial concessions, it does provide a link between the historic disenfranchisement I have outlined in this chapter and the potential futures still to come. Chapter 6 offers a deeper exploration of food-focused visitor services that recenter Indigenous foodways within national parks. As the National Congress of American Indians has called on the US federal government and its constituent agencies "to protect, enhance, restore and assure tribal access to the First Foods" (Brigham 2014)—traditional foods eaten precontact—the time is ripe for a reckoning around foodways on park-lands comprising Indigenous ancestral lands. To understand the context of these contemporary efforts, it is necessary to examine the present-day national parks foodscape more broadly and to understand the regulatory environment and the role of media in shaping and responding to cosmopolitan desires.

Notes

1 A note on language: when referring collectively to multiple groups of people with preex-isting sovereignty prior to settler contact on the land now occupied by the United States, I use, and cite sources that use, "Indigenous," "Native American," or "American Indian" interchangeably, while recognizing that there is no naming convention favored equally by all Indigenous groups or that escapes controversy. When referring to members of a single community or nation, I use the specific group name (e.g., Havasupai, Cheyenne, Zuni). For more on this topic, see Michael Yellow Bird, "What We Want to Be Called: Indigenous Peoples' Perspectives on Racial and Ethnic Identity Labels," *American Indian Quarterly* 23.2 (Spring 1999): 1–21.
2 According to Bingham (2019), Mackinac National Park contained a military installa-tion, Fort Mackinac, and was administered by the War Department in the years prior to the founding of the NPS. After the federal government decided that the Great Lakes location was not a strategic military priority, it decommissioned the fort and relocated the troops, leaving no one to administer the park. It was handed over to the State of Michigan in 1895 and became a state park.
3 Lest I paint too monolithic an NPS perspective on this issue, I'll point out that there was some empathy within the NPS for the issue of Indigenous subsistence. NPS biologist George Wright wrote in 1933, "Certainly the Indians who face starvation with the com-ing of very winter are not to be condemned if they step across the line and take for the gratification of hunger what the white man tries to conserve for the satisfaction of aes-thetic longings" (George Wright et al., *Fauna of the National Parks of the United States.* Washington, DC: GPO 1933, p. 94, cited in Burnham 110), yet federal and state laws and park policy nonetheless aligned for over a century against Indigenous usufructuary rights.
4 I find the claim in scholarly literature that farming/agriculture was entirely foreign to Indigenous peoples in what is now the US to be overrepresented. Indeed, many Ameri-can Indian communities, particularly in the East, relied on agriculture and the raising of crops long before settler contact. For more on this, see Dana Vantrease, 2013, "Commod Bods and Frybread Power: Government Food Aid in American Indian Culture." *The Journal of American Folklore* 126, 499 (Winter): 55–69, or Hoover and Mihesuah 2019.

5 According to White, the fixed-harnessed plow was put into use in Northern Europe, where the soil was stickier than in the Near East and Mediterranean, during the seventh century BC. Early plows were pulled by just two oxen and simply scratched the soil, whereas the fixed-harnessed plow turned the soil and required eight oxen to pull it. The tilling capacity of the new plows altered the way land was distributed: No longer was subsistence farming on small squares of land to support families possible—instead, peasants had to pool their oxen, working long strips of land in a violent and extractive fashion (White 1967, 1205).

References

Akerman, James R. 2020. "Science, Wonder, and Tourism in the Early Mapping of Yellowstone National Park." In *Cartographic Expeditions and Visual Culture in the Nineteenth-Century Americas*, edited by Ernesto Capello and Julia B. Rosenbaum, 167–97. New York: Taylor and Francis.

Bartlett, Richard A. 1985. *Yellowstone: A Wilderness Besieged*. Tucson: University of Arizona Press.

Bingham, Emily. 2019. "Mackinac Island Used to be America's Second National Park." *Mlive.com*, March 2. https://www.mlive.com/life-and-culture/g66l-2019/03/e5f658dfa63930/mackinac-island-used-to-be-americas-second-national-park.html.

Blodgett, Peter J. 1990. "Visiting 'The Realm of Wonder': Yosemite and the Business of Tourism, 1855–1916." *California History* 69 (2): 118–33.

Brigham, Kathryn. 2014. "Urging the Federal Government to Safeguard Tribal First Foods." Resolution ATL-14-022 passed by the National Congress of American Indians, Atlanta, GA. https://www.ncai.org/resources/resolutions/urging-the-federal-government-to-safeguard-tribal-first-foods.

Burnham, Philip. 2013. *Indian Country, God's Country: Native Americans and the National Parks*. Bloomington, IN: Authors Guild/iUniverse.

Clayton, John. 2020. "Who Gets to Hunt Wyoming's Elk? Tribal Hunting Rights, U.S. Law, and the Bannock 'War' of 1895." *WyoHistory.org*, September 29. https://www.wyohistory.org/encyclopedia/who-gets-hunt-wyomings-elk-tribal-hunting-rights-us-law-and-bannock-war-1895.

"Declaration of Nyéléni." 2007. Sélingué, Mali. February 27. chrome-extension://efaidn-bmnnnibpcajpcglclefindmkaj/https://nyeleni.org/IMG/pdf/DeclNyeleni-en.pdf.

Dunnell, Mark H. 1872. "The Yellowstone Park. Report to Congress from the Committee on Public Lands, to Accompany Bill H. R. 764." Presented February 27. https://www.yellowstone.co/history/congress/congress15.htm.

Grove, Richard H. 1995. *Green Imperialism: Colonial Expansion, Tropical Island Edens, and the Origins of Environmentalism, 1600–1800*. Cambridge: Cambridge University Press.

Grusin, Richard. 2004. *Culture, Technology, and the Creation of America's National Parks*. Cambridge: Cambridge University Press.

Hall, C. Michael and Warwick Frost. 2009. "National Parks and the 'Worthless Lands Hypothesis' Revisited." In *Tourism and National Parks: Perspectives on Development, Histories, and Change*, edited by Warwick Frost and Michael C. Hall, 45–62. New York: Routledge.

Hennessy, Maggie. 2021. "Interior Secretary Deb Haaland Cooks to Share Her Indigenous Heritage." *Food & Wine*, June 23. https://www.foodandwine.com/news/interior-secretary-deb-haaland-cooking.

Hoover, Elizabeth and Devon A. Mihesuah. 2019. "Introduction." In *Indigenous Food Sovereignty in the United States: Restoring Cultural Knowledge, Protecting Environments, and Regaining Health*, edited by Devon A. Mihesuah and Elizabeth Hoover, 3–25. Norman: University of Oklahoma Press.

Hufford, Mary. 1986. *One Space, Many Places: Folklife and Land Use in New Jersey's Pinelands National Reserve*. Washington, DC: American Folklife Center.

Hurt, Douglas R. 1987. *Indian Agriculture in America*. Lawrence: University Press of Kansas.

John Milner Associates. 2005. "Cedar Pass Developed Area Badlands National Park Cultural Landscape Report." Prepared for the National Park Service. https://www.nps.gov/parkhistory/online_books/badl/cedar_pass_clr.pdf.

Kantor, Isaac. 2007. "Ethnic Cleansing and America's Creation of National Parks." *Public Land and Resources Law Review* 28 (5): 41–64. https://scholarworks.umt.edu/cgi/viewcontent.cgi?article=1267&context=plrlr.

Keller, Robert H. and Michael F. Turek. 1998. *American Indians and National Parks*. Tucson: University of Arizona Press.

Lewis, Courtney. 2012. "The Case of the Wild Onions: The Impact of Ramps on Cherokee Rights." *Southern Cultures* 18 (2): 104–17.

Lindholm, Melanie M. 2019. "Alaska Native Perceptions of Food, Health, and Community Well-Being." In *Indigenous Food Sovereignty in the United States: Restoring Cultural Knowledge, Protecting Environments, and Regaining Health*, edited by Devon A. Mihesuah and Elizabeth Hoover, 155–72. Norman: University of Oklahoma Press.

Mark, Stephen R. and C. Michael Hall. 2009. "John Muir and William Gladstone Steel: Activists and the establishment of Yosemite and Crater Lake." In *Tourism and National Parks: Perspectives on Development, Histories, and Change*, edited by Warwick Frost and Michael C. Hall, 88–101. New York: Routledge.

Meeker, Joseph W. 1973. "Red, White, and Black in the National Parks." *The North American Review* 258 (3): 3–7.

Mertens, Richard. 2021. "Seeds and Beyond: Native Americans Embrace 'Food Sovereignty.'" *The Christian Science Monitor*, February 22. https://www.csmonitor.com/USA/Society/2021/0222/Seeds-and-beyond-Native-Americans-embrace-food-sovereignty.

Muir, John. 1890. "The Treasures of the Yosemite." *The Century Magazine*, 40 (4). Accessed February 27, 2021. http://www.yosemite.ca.us/john_muir_writings/the_treasures_of_the_yosemite/.

Muir, John. 1911. *My First Summer in the Sierra*. Boston, MA: Houghton Mifflin. Last updated February 3, 2020. https://www.gutenberg.org/files/32540/32540-h/32540-h.htm.

National Park Service. 2023. "Tribal Plant Gathering in National Parks Tribal Consultation." Posted June 23, 2023. https://www.nps.gov/articles/000/tribal-plant-gathering-consultation.htm.

Olmsted, Frederick Law. [1865] 1994. "The Yosemite Valley and the Mariposa Big Tree Grove: Olmstead Report on Management of Yosemite, 1865." In *America's National Park System: The Critical Documents*, edited by Lary M. Dilsaver, 6–19. Lanham, MD: Rowman and Littlefield. https://www.nps.gov/parkhistory/online_books/anps/anps_1b.htm.

Parasecoli, Fabio. 2017. "Cosmopolitan Foodies and Local Food." *The Huffington Post*, December 4. https://www.huffpost.com/entry/cosmopolitan-foodies-and-local-food_b_5a2534cee4b04dacbc9bd8e6.

Patterson, Allie. 2022. "Indian Removal from Yellowstone National Park." *Intermountain Histories*, July 27. https://www.intermountainhistories.org/items/show/344.

Pennington, Emily. 2022. "Deb Haaland, Tracy Stone-Manning, and Charles 'Chuck' Sams III Are Trying to Fix the Mess Trump Left Behind." *Outside*. December 7. https://www.outsideonline.com/outdoor-adventure/environment/deb-haaland-tracy-stone-manning-charles-chuck-sams-iii-outsiders-2022/.

Pratt, Mary Louise. 1992. *Imperial Eyes: Travel Writing and Transculturation*. New York: Routledge.

Rice, Stian. 2022. "Divide and Cultivate: The Role of Prisons and Indian Reservations in U.S. Agricultural Imperialism." *Food and Foodways* 30 (1–2): 16–37. https://doi.org/10.1080/07409710.2022.2030935.

Rothman, Hal K. 1998. *Devil's Bargains: Tourism in the 20th Century American West*. Lawrence: University of Kansas Press.

Ruppert, David. 2003. "Building Partnerships between American Indian Tribes and the National Park Service." *Ecological Restoration* 21 (4): 261–3.

Sams III, Charles F. 2022. "Tribal Co-Management of Federal Lands: Acknowledging the History and Considering the Path Forward." Statement of Director of the National Park Service, US Department of the Interior, before the House Committee on Natural Resources, Regarding Tribal Co-Management of Federal Lands. Office of Congressional and Legislative Affairs. March 8. https://www.doi.gov/ocl/tribal-co-management-federal-lands.

Scott, David and Kay Scott. 2019. "National Parks 101: Who Runs the Lodges and Dining Establishments?" *NationalParksTraveler.org*, February 27. https://www.nationalparkstraveler.org/2019/02/national-parks-101-who-runs-lodges-and-dining-establishments.

Spence, Mark. 1999. *Dispossessing the Wilderness: Indian Removal and the Making of the National Parks*. New York: Oxford University Press.

"The Hostile Savages." 1872. *The New York Times*, September 10: 1. https://timesmachine.nytimes.com/timesmachine/1872/09/10/79190807.html?pageNumber=1.

US Department of State. 1868. "Fort Laramie Treaty." *U.S. Statutes at Large* 15: 635–48.

US Department of the Interior. 2021. "Interior and Agriculture Departments Take Action to Strengthen Tribal Co-Stewardship of Public Lands and Waters." Posted November 15. https://www.doi.gov/pressreleases/interior-and-agriculture-departments-take-action-strengthen-tribal-co-stewardship.

White, Jr., Lynn. 1967. "The Historical Roots of Our Ecological Crisis." *Science* 155 (3767): 1203–7.

Wolfe, Patrick. 2006. "Settler Colonialism and the Elimination of the Native." *Journal of Genocide Research* 8 (4): 387–409.

Wolfley, Jeanette. 2016. "Reclaiming a Presence in Ancestral Lands: The Return of Native Peoples to the National Parks." *Natural Resources Journal* 56 (1): 55–80.

Zeppel, Heather. 2009. "National Parks as Cultural Landscapes: Indigenous Peoples, Conservation, and Tourism." In *Tourism and National Parks: Perspectives on Development, Histories, and Change*, edited by Warwick Frost and C. Michael Hall, 259–81. New York: Routledge.

4

SWALLOWING TENSIONS

Exploring the Contemporary Foodscape

Early in 2018, more than 25,000 people signed a Change.org petition to "Stop Starbucks in Yosemite," apparently agreeing with the sentiment of its creator that "multinational corporations have no place in our national parks" (Concerned Citizen 2018). Park lovers were outraged that Yosemite Valley Lodge, operated by multinational giant Aramark, was planning to open a Starbucks kiosk in its Basecamp Eatery food court. Although some fretted about a possible increase in litter or said they thought it would be better to support small businesses, a great number of the posters articulated their reasons for signing the petition along the line of one signer, Terry Barber: "Bringing corporate giants into a National Park may be the beginning of the end for an iconic American place." What many of the petition signers seemed not to realize is that US national parks are already foodscapes dominated by large corporations.

This chapter examines the foodscape of today's national parks and investigates how the parks and their approved concessioners narrativize these foodscapes and the identity of those who patronize them. How are the competing mandates of US national parks—preservation of natural spaces and enjoyment of these spaces by the citizenry—manifest or reconciled within the possibilities for eating in the parks? What does the nation's reliance on industrial food concessions in the parks reveal about the tensions inherent in the mission of the park service? To answer these questions, I provide an overview of the US national park foodscape and a more detailed analysis of the way eating is narrativized by the dominant concessioner in Yellowstone National Park as a case study.

From the Latin *concession-*, from the verb *concedere*, "concession" means a yield or grant, and concessions have come in the US to be associated with food or snack stands, sold in a venue that does not operate the stands; the venue has made a concession to a vendor, a concession to the public. It's a seemingly "meh"

DOI: 10.4324/9781003455516-5

arrangement, rooted in compromise from the start. The venue says, "Okay, we'll sell you something to eat, because you're a human and you're going to get hungry, but food is not the main point of why you are here." And the public says, "Okay, I'll accept the limited choices and perhaps lower quality, because at least there's something to eat here, and it's awfully far or inconvenient to eat elsewhere."

Currently, the National Park Service (NPS) manages 425 individual units in the 50 states, Washington DC, and US territories; these include 63 official national parks, as well as a host of other types of designations, including national monuments, national historic sites, national historical parks, national recreation areas, national battlefields, national seashores, and the list goes on (National Park Service 2022b). Although I occasionally mention these alternative kinds of units in this book, the foodscapes of the 63 official parks are my primary concern, not only because this helps me to manage scope but also because the official parks have an outsized and special role in the American public imaginary.

Visitors to most parks have limited eating options. Some choose to eat in restaurants outside the parks in gateway towns, which are often well set up to accommodate hungry tourists. Others pack a cooler and take advantage of picnicking opportunities, happy when they can find an open picnic table, or they tote a blanket to spread under a shady tree. Some camp and enjoy cooking their own food over the campfire. But for most visitors to the parks—those both unprepared with their own food and those hoping for a tasty or unique or possibly just convenient nosh inside park borders—the dining options are those provided by concessioners.

What is a park food concession? These days, they are most commonly grills, snack bars, or cafeteria-style eateries, serving a basic menu meant to appeal broadly. Their food is generally not fancy, and the emphasis is on scale, with operations meant to serve a large number of park visitors in as efficient a fashion as possible. Stalwart dishes like burgers, pizza, chicken tenders, and salads appear on park menu boards across the land from Denali National Park in Alaska to Everglades National Park in Florida. To win concessions contracts, the businesses that operate the venues must ensure that their prices are relatively affordable, and that the food meets specified nutritional requirements, along with the predictable health and safety practices. A small number of concessioners operate more aspirational restaurants—sit-down venues with table service, long waits in high season, a more curated menu, and (often) priceless views of some of the country's most remarkable landscapes, like the fiery caldera of Hawaii Volcanoes National Park and the stunning peaks of Grand Teton National Park.

According to NPS guidelines, concessioners are not only responsible for providing high-quality facilities and services, but they are also required to participate in the interpretation of the park to visitors. This interpretive mission can be accomplished in any number of ways, according to the NPS: "through guided activities; the design, architecture, landscape, and décor of facilities; educational programs; interpretive menu design and menu offerings; and involvement in the park's overall interpretive program" (National Park Service 2006, 145). The park service takes an active role in evaluating the concessions to ensure that their rates are fair, that they are complying with best environmental management practices, and, for food service operations, that their compliance with health and sanitation standards is up to snuff.

Ownership and operation of park concessions have shifted over the years, from a somewhat chaotic scene with independently owned, hypercompetitive business in the early years of the parks (1872–1915), to a consolidated "one park, one concessioner" model in the decades after the 1916 establishment of the NPS. National park concessions today are quasi-monopolies that tend to be run by conglomerates. Large companies win government contracts for exclusive rights to provide food, lodging, gift shops, and other activities and services in the park. A few behemoths hold contracts in multiple parks, raking in a very large share of overall concessions income.

Currently, four large companies—Aramark, Delaware North, Ortega, and Xanterra—operate the lion's share of park food concessions in 36 parks that currently have them (National Park Service 2022a).

TABLE 4.1 Major National Parks Concessioners (as of 2023)

Aramark	*Ortega*
Crater Lake	Acadia
Denali	Death Valley
Glacier Bay	Hawaii Volcanoes
Mesa Verde	Carlsbad Caverns
Olympic	White Sands
Yosemite	Petrified Forest
Badlands	Lassen Volcanic
Big Bend	Mammoth Cave[a]
Bryce Canyon	Great Smoky Mountains
Grand Teton	
Isle Royale	
Grand Canyon North	
	Xanterra
	Crater Lake
Delaware North	Death Valley[b]
Grand Canyon	Glacier
King's Canyon	Grand Canyon
Olympic	Rocky Mountain
Sequoia	Shenandoah
Shenandoah	Yellowstone
Yellowstone	Zion

[a] Not listed in the NPS directory but shows on Mammoth Cave Lodge website as Ortega and on MCNPS website.
[b] Furnace Creek Resort on "inholding" private land that is surrounded by NP; not an officially contracted NPS concession.

The quasi-monopolistic nature of these contracts means that in many parks, one single company operates every food service vendor within the park. For instance, at Mesa Verde, Aramark is the only game in town, operating the cafeteria-style Spruce Tree Terrace Café, the food court-style Far View Terrace Café, the fancier fine-dining Metate Room, the sports-pub Far View Lounge, the campground-based breakfast-oriented Knife Edge Café, and even the Morefield Campground Store, for one's snacking needs. In other parks, there might be a few concessioners present, like at Yellowstone, where Xanterra runs about 20 restaurants and where Delaware North has a smaller contract to operate a dozen general stores that also sell food. The domination of the park landscape by a handful of companies means a uniform and predictable, not particularly distinctive, experience for travelers. These large companies covet concession contracts because of their profitability, as they are uniquely protected from competition under favorable terms of contract and were historically granted preferential rights to contract renewal (Hartzog 1988).

The remainder of the food concessions in the parks are run by subsidiaries of larger companies whose primary business is not in national parks, smaller companies not focused on food, or, rarely, an individual proprietor. In the subsidiary category, at Mount Rainier, visitors can eat in a handful of cafes or lodge dining rooms, all operated by Rainier Guest Services, a subsidiary of Guest Services Inc, which also operates at the National Mall and Stehekin in North Cascades. Or at Gateway Arch, the food concessions are run by Arch Café, LLC, an Evelyn Hill Company, which also operates concessions at Statue of Liberty and Ellis Island. In the category of smaller companies not focused on food, a handful of food concession contracts are with businesses that primarily operate transit concessions but also vend food to customers while they are traveling within or to a park; for instance, Island Packers Corporation (Channel Islands) and Yankee Freedom III (Dry Tortugas) are one-park operations primarily focused on boating with snack bars on board. Very, very few park concessions are of the "mom and pop" variety, given both economies of scale and the complex bidding and contracting process, but a handful do exist, like the privately managed Great Basin Café in Nevada's Great Basin National Park.

A few parks have multiple food service concessioners. For instance, the NPS has established food service contracts in Everglades National Park with four different small outfits (Coopertown Everglades Airboat Tour and Restaurant, Everglades Safari Park, Gator Park; and Everglades Guest Services LLC); likewise, there are three companies (Olympic Peninsula Hospitality LLC alongside biggies Delaware North and Aramark) providing food service operations at Olympic National Park. But there are no food concessions at all in nearly half of the nation's 63 official national parks: 27 of the parks, including some heavily visited sites like Arches and Joshua Tree, and some truly remote ones like Lake Clark and American Samoa, have no food services onsite whatsoever.[1]

Even when there's no official concession contracted with the NPS, visitors may sometimes still be able to access food. For instance, at Indiana Dunes National

Park, the Dunes Pavilion is situated on state park land, not national park land, but this would be imperceptible to the national park visitor who strolls up from the sand for some soft serve. Likewise, most visitors to the eatery at the posh Inn at Death Valley will not realize that they are on an inholding, a privately owned piece of land surrounded by Death Valley National Park, and thus what's on offer there is not subject to the same requirements as official park concessions. (Being run by Xanterra means that the offerings will be quite familiar anyhow.) Nor might eaters at the Superior Bathhouse Brewery or Eden, the Hotel Hale's restaurant, realize that they are in Hot Springs National Park but not eating at a park-authorized concession—the only giveaways are the variety, price, and foodie-ness that are typically not found in an official concession.

Park concessions are profitable for corporate managers, with contracts guaranteeing the conditions for wider margins than the restaurant business in general. The market is literally hungry; according to an NPS report on visitor spending effects, in 2021, park visitors spent $4.2 billion at restaurants and bars and $1.5 billion on food at grocery and convenience stores (National Park Service 2021, 12). Food concessions at Yellowstone National Park alone, for instance, generated over $30 million in revenue in 2016 (*Government Food Service* 2017). The economies of scale that allow large companies to employ staff solely for the purposes of contract procurement and to install cookie-cutter operations in multiple parks make business sense, and result in a standard and predictable eating experience at most parks.

Beyond the highly visible NPS-authorized park concessions, eateries in gateway towns, and whatever day-use visitors may bring for a picnic or trail snack, or what campers may tote to eat at their campsites, a small subset of national parks today permit a surprising foodscape element: hunting, fishing, and foraging. Sixty-six areas managed by the NPS allow recreational hunting; seven Alaskan areas permit federal subsistence hunting through the Alaska National Interest Lands Conservation Act (ANILCA); Badlands National Park sanctions Tribal hunting; and one part of Grand Teton National Park, coordinating with the State of Wyoming, permits controlled elk reductions (US Department of the Interior 2017). Hunting, fishing, and foraging have had a fraught history on parklands, particularly for Indigenous people, a theme I treat in more depth in other chapters. Here, I focus my examination on the most visible park foodways, concessions.

The style of concessions that most park travelers today are familiar with originated during the Depression. The rustic and elegant grand lodges for which many parks are famous—including the Grand Canyon's El Tovar (1905), Old Faithful Inn (1903–04) at Yellowstone, the Many Glacier Hotel (1915) at Glacier National Park, and Crater Lake Lodge (1915)—were built in the first two decades of the twentieth century. They catered to wealthy tourists who arrived at the parks by train, and they made steep profits by providing sightseeing trips around the parks to these travelers and by serving them in the grand dining rooms of the lodges. In the 1920s, the mass production of the automobile made travel affordable and accessible to a wider range of people, and middle-class tourists began to drive themselves to the parks. They didn't need the sightseeing transportation of their wealthier predecessors, but they did need faster food in the far corners of the parks to which they could now travel independently. Hence, in the 1930s, concessioners began to invest in more cafeterias, grills, coffee shops, gift shops, and cabins to serve the growing number of visitors (Mantell 1979, 16).

The regulation of national park concessions has waxed and waned over the nearly 100 years since they started popping up like mushrooms after a rain. Park visitations climbed dramatically through the middle of the twentieth century, from 120,690 recreation visits in 1904, to 1,900,499 in 1925, to a whopping 32,706,172 in 1950, as the number of parks grew, and more and more Americans and international visitors made their way (National Park Service 2022c). Concessioners had great difficulty keeping up with demand, and the quality of their offerings declined, resulting in a period of strict regulation after World War II. Deterred by the hassles and expense of such regulation, many concessioners let their contracts expire, leaving the Department of the Interior unable to find new concessioners (Mantell 1979, 18n113). An agreement in 1950 and the later 1965 Concessions Policy Act provided a framework for concessions management, safeguarding concessioner investments and protecting the public from inflated prices and sub-par facilities. The 1965 Act helped to line the pockets of concessioners, insuring them against loss and underwriting their profit-generating capital investments within the park, and giving them preferable rights of renewal of their long-term quasi-monopoly contracts (Herman 1992, 45).

Critics have noted the role of concessioners in attracting visitors to the parks.

A concessioner cannot build a 500-room luxury hotel in the middle of a national park, hire a five-star gourmet chef to run the kitchen, promote it with slick color ads in the trendiest magazines, and then claim to be blameless when 1,000 people show up to stay there every night in the summer. Concessioners create demand, and they must be held accountable for its effects.

(Herman 1992, 37)

Seen in this light, the lackluster food of the national parks may be a boon for environmental preservation, as very few people are heading to the parks for the eats, at present. Still, even with limited foodie draw, national park foodways, largely circumscribed by corporate concessioners, shape and are shaped by projects of self-making that happen in the guise of romantic consumption. A case study of the concessions in America's oldest national park promises to illuminate this phenomenon, but first, a question: who is park food for?

Catering to Whom?

From the get-go, national parks were conceptualized in conflicting ways with regard to access. Although the parks were touted as spaces for everyone, early proposals for a park bureau framed the parks as well-managed resorts—"recreational spaces and manipulated gardens overseen by guardians of taste and civility" (Taylor 2016, 330). Key figures associated with the early NPS, including Harold Bryant, Madison Grant, and Charles Goethe, were prominent eugenicists who supported restrictive immigration measures. Goethe in particular supported park naturalist programs, hoping that they would make people more aware of biological selection processes and thus more reflective on their own breeding practices, and thus likely to politically support strict immigration laws (Taylor 2016, 347). Preservationism and conservationism were intertwined with racist nativist and eugenicist discourse that reverberates still today in our understandings of whose souls the parks are meant to feed.

As Meeker pointed out nearly 50 years ago, the parks have not succeeded in appealing equally to all people: "Poor people, black people, and ethnic minorities generally show little enthusiasm for the park idea … The parks stubbornly remain essentially playgrounds for middle-class citizens" (Meeker 1973, 5). Indeed, visitation rates remain low for much of the US population, among particularly people of color and the socioeconomically marginalized (Weber and Sultana 2013). People who have been denigrated for their connections to the land might not just be indifferent—they may be antagonistic to national parks and their self-making possibilities, better reserved for a white middle class striving toward cosmopolitanism. Some parks are easily accessible to US population centers, but a great many are remote, and require a good amount of free time and significant expense to reach (Bartlett 1985). Yet there exist important efforts to connect people of color with the parks, to ensure that the transformative personal experiences that public lands offer are not exclusive; this happens both by increasing accessibility measures within the parks and by raising awareness and spurring interest within communities of color (Peterman and Peterman 2009).

Presently, the food offerings in US national parks are mass oriented. As the analysis in the next section shows, there are glimmers of aspiration toward feeding a savvier, foodie-informed audience, but that is less so because of the promised revenue from this niche cosmopolitan market and more because the cosmopolitan niche exerts significant symbolic power on the mass market. Still, the concessions system is primarily oriented to scale, focused on feeding lots of visitors in the most efficient way possible that can generate revenue for the companies that operate the concessions. The federal government ensures, in the process of awarding concession contracts, that the prices are reasonable and that healthy offerings are in the mix. But in the big picture, food is not used in the parks to attract a more ethnically, racially, or internationally diverse crowd of visitors, and there is little allowance for religious or cultural difference.

Yellowstone and Xanterra: How to Eat in the Park and How to Be

US national parks are not known for the opportunities to eat that they provide, and the NPS does little to emphasize the park dining experience. However, food concessioners are eager storytellers, and many of the largest have become adept at using digital media to communicate about their role in the park foodscape. I am interested in how competing park mandates for preservation of natural spaces and enjoyment by the public are manifested and reconciled by concessioners in the stories they tell about themselves. There are dozens of concessioners operating hundreds of food venues in dozens of parks, so to say something generic about their storytelling is to miss the juiciness of the specific. Thus, I turn my attention in this chapter to a case study of how to eat in one park, the nation's first and one of the most lucrative parks for concessions contracts, Yellowstone.

Hospitality companies, eager to attract visitors from across the US and beyond, rely on their websites and on digital marketing to reach their geographically dispersed audiences (George 2021). Many park visitors find their way to the hospitality companies that run park concessions through the park-specific websites of the NPS. Likely to the chagrin of these concessioners, restaurants do not occupy a particularly prominent place on the websites of any of the US national parks, which are designed in a standard fashion for ease of navigation. From the home page of any park, one needs to navigate to "Plan Your Visit," then click "Eating and Sleeping." Eating is never listed under "Things to Do," as an important activity, like hiking, photographing, attending a ranger talk, or watching wildlife. Nor are the restaurants easily navigable from the prominently featured "Places to Go" page, which instead features the different areas of the park and then offers information about key landscape attractions, outdoorsy activities, and answers to frequently asked questions. From "Eating and Sleeping," one chooses "Lodging, Camping, Picnicking or Restaurants" on the Yellowstone website.

The picnicking page features a rather unhappy looking group of white picnickers—possibly a family—sitting at a picnic table with the stately sandstone buildings of the Mammoth Hot Springs in the Fort Yellowstone Historic District visible in the background. They have an open bag of potato chips, and one can make out a few plastic containers, mayo, and catsup. There are a few reusable water bottles, too. The text on the site is mostly informative, noting that there are several sites for picnicking in the park, but the chief message concerns safety: would-be picnickers are warned to refrain from feeding animals, to store food properly, and to use stoves and fire responsibly. There's also a map of 52 picnic spots in the park, and a listing of the locations of over 300 picnic tables. There's nothing about conviviality, authenticity, or solitude; as one would expect of the NPS, they're not exactly "selling" the non-commodified picnic experience, but merely providing information. Lower on the page, one can click through to various concession-based eating and lodging experiences.

The "Restaurants" link on the Yellowstone site is a more robust portal that takes the reader to a comprehensive (yet unadorned by photos or even adjectives) listing of summertime dining options within the park's boundaries, including fine-dining restaurants, cafeterias, snack bars and coffee counters, ice cream stands, and bars. Clicking any of the linked restaurants takes one outside of the NPS site to any number of pages run by Xanterra, one of the largest authorized concessioners in the NP system. (Although its general stores are listed on the restaurant page, there are no links provided to any of these enterprises run by Delaware North, a chief Xanterra competitor). Clicking through to the world of Xanterra means moving from the factual and straightforward information presentation of the NPS to a more engaging and persuasive space where opportunities abound for would-be eaters to imagine themselves.

The Xanterra Travel Collection runs concessions in Yellowstone, Grand Canyon, Glacier, Zion, and Rocky Mountain National Parks, as well as at Mount Rushmore National Memorial. They operate the historic Oasis at Death Valley National Park, but on an inholding of private land, so they're not an official concessioner there. Xanterra operates other park-adjacent concessions, including the Grand Canyon Railway and Hotel and the Grand Hotel just outside the Grand Canyon, and Cedar Creek Lodge flanking Glacier National Park. Apart from the park system, they also operate Windstar Cruises, VBT Bicycling Vacations, Country Walkers, and Holiday Vacations. This portfolio, which allows them to post nearly a billion dollars annually in top-line revenue (Weissman, n.d.), indicates Xanterra's aptitude at selling vacation dreams. Xanterra is owned by the massive, highly diversified Anschutz Corporation, an American private holding company that also owns oil fields, mines, arenas, stadiums, TV stations, newspapers, and sports teams. Today, Xanterra is descended from The Fred Harvey Company, which operated the first restaurant chain in the US in the 1870s. The Harvey Company was known for its choice restaurants and hotels situated alongside railways in the American Southwest (Xanterra 2022d). Xanterra has the main concessions contract at Yellowstone until 2033, having won a 20-year term in 2013.

Xanterra makes about a third of its revenue from food and beverage operations (Blevins 2013, np) and is cited as one of the "most environmentally aware tourist operations in the country, with large-scale renewable energy projects and zero-waste goals" (Blevins 2013, np). Andrew Todd, CEO of Xanterra, describes his company's environmental record as "unrivaled"—"We are definitely not into corporate greenwash" (cited in Blevins 2013). Xanterra's operation at Yellowstone is a financial success, generating over $87 million in gross receipts in 2011 alone (Miller 2017).

Who one can be when one eats at a Xanterra property, according to Xanterra's representational rhetoric, aligns with the founding rhetoric of the national parks, as discussed earlier, as well as the tenets of contemporary cosmopolitanism. Xanterra beckons to tourists with a taste for distinction:

> For conscious travelers and explorers who want to satisfy their longing for enrichment, Xanterra Travel Collection provides intimate, immersive experiences around the world. Xanterra Travel Collection has been operating in legendary destinations for over 150 years, providing up close access and uncommon expertise for truly exclusive, unforgettable experiences, and we do so while treading carefully with a softer footprint and being mindful of our responsibility to others and to the earth.
>
> *(Xanterra 2022a)*

How these promises of enrichment, access, immersive experience, careful treading, and responsibility play out across Xanterra's properties in Yellowstone is worth a close examination. After visiting dozens of their eateries and extensively reviewing their website, I can describe Xanterra's food zeitgeist in five words that appear frequently throughout their narratives: fresh, local, house-made, customizable, and sustainable. As they are beholden to offer food at a variety of price points, I analyze how Xanterra's storytelling about a basic eating venue, a higher-end, more expensive restaurant, and a specialty dining experience serves to reinforce ideas about romantic consumption and the character of the eater.

"Customize Your Meal with Fresh Ingredients": Canyon Lodge Eatery

The basic Xanterra dining experience is represented by Canyon Lodge Eatery, a casual cafeteria in a visually appealing midcentury modern building with a Googie/Populuxe/Doo Wop design vibe—all soaring ceilings, space-age starburst light fixtures painted blue, green, and tangerine, and tables, chairs, and umbrellas in a rainbow of futuristic hues. It is situated in Canyon Village, a sprawling area featuring the greatest concentration of accommodations in Yellowstone, on the east side of the park near the Grand Canyon of the Yellowstone River. Unlike some of the park's other eateries, for instance, those overlooking the iconic Old Faithful geyser, the restaurant does not have much of a view, unless one likes parking lots. Canyon Lodge Eatery is a busy place: It might serve over 450,000 meals in a typical May-to-September season, with 3,000–3,900 transactions a day. But for the most part, the cafeteria is not pricey; in 2019, the average check was $10.36. Total seasonal food sales for the 149 days they were open were $4,768,444, plus another $406,305 for beverages (Boss 2019).

The way Xanterra communicates about the dining experience at Canyon Lodge Eatery has changed, presumably only temporarily, during the COVID years. Lots of evocative details and photos that were present on the restaurant's web page in 2020 have subsequently disappeared, apparently in response to supply chain issues and labor shortages that have made promising too much a fool's errand. Whereas in 2022, the restaurant's website directs eaters to visit the restaurant itself to get ahold of a menu, its 2020 and earlier iteration emphasized the transformative power of customization, appealing to eaters aspiring to control their own destinies: With an establishing photo that is a wide shot of the restaurant's mostly empty interior, depicting a smattering of white families eating on colorful barstools in the Doo Wop style of diners everywhere, website visitors were hailed with the phrase, "Customize Your Meal with Fresh Ingredients."

The Canyon Lodge Eatery website detailed two cafeteria-style service bar options for customizing lunch or dinner with "fresh sustainable items, including local ingredients as available"—language that is suggestive of Xanterra's ambitions, but not overly committal: "Fresh Woks" and "Slow Food Fast." At the Fresh Woks counter, what customers can select has changed over the years, presumably in response to supply chain issues, but generally they are invited to choose a rice or noodle base, a protein, a sauce, and some toppings. Prior to the complicating factor of supply chain issues, various versions of the Fresh Woks menu nodded toward the local and sustainable. In one version, from 2020, the only food item on the menu that indicated anything about provenance was the Wild Alaska Pollock, which might be considered local in Kenai Fjords National Park but is a far cry from local to northwestern Wyoming. From $7.95 for kids to $13.95 for the most jazzed-up regular wok, the 2020 prices are consistent with national norms for fast casual outside the park. The only food item on the menu labeled as locally sourced is an "Elliott's of Montana Fresh Baked Cookie." The menu does contain a key for which items are vegetarian (V), vegan (VG), or gluten-free (GF), but its creators seem to have forgotten to label most of the items. There is a special green leaf icon for "menu items made with sustainable and/or organic ingredients," but only the Alaskan pollock earns this designation on the Fresh Woks menu.

Beyond Fresh Woks, the other option at Canyon Lodge Eatery is "Slow Food Fast," framed as another "choose own adventure" kind of setup, this time sans wok. The same menu key exists here, but not a single thing has a V, VG, GF, or the vaunted green leaf, despite listings that include "Organic Chicken and Vegetable Chili" and "Beyond Meat™ Plant-Based Swiss Steak." Xanterra knows that sustainability, locality, and customization are appealing to today's cosmopolitan travelers, and they nod to these concepts on their website and menu. But Xanterra provides little of the real information that would allow eaters to facilitate the purchase of sustainable and local foods.

The breakfast menu at Canyon Lodge Eatery does go somewhat further in signifying local and sustainable options, noting the availability of local hot cereals (Cream of the West™ Whole Grain or Montana Milling™ Oatmeal), vanilla almond milk with no genetically modified organisms (GMO), Rainforest Alliance coffee, Blue Sky™ non-GMO organic sodas, and organic flavored sparkling water, as well as affordable reusable cups that will allow refills at a discount. Not designated as organic or sustainable, but surely of interest to today's demanding diner, are the Kevita sparkling probiotic and Kevita kombucha, signaling that we're not at your dad's national park canteen anymore.

Although breakfast is more promising, the enrichment-seeking, conscious traveler hailed on Xanterra's homepage is likely to be disappointed in lunch or dinner at the Canyon Lodge Eatery—a phenomenon borne out in TripAdvisor reviews, where (as of this writing) it comes in dead last, #27 out of the 27 Yellowstone eateries reviewed on the site. (I discuss reviews in more depth in Chapter 5.) But perhaps, at the lower price point that prevails at Canyon Lodge Eatery, despite the trendy rhetoric of choice and customization, the cosmopolitan visitor is not at all the target audience.

"Where Delicious Meals Can Be Enjoyed": Canyon Lodge M66 Grill

In contrast to basic eateries like the Canyon Lodge Eatery, mid-range dining experiences in Yellowstone tend to provide more of the concrete information sought by cosmopolitan travelers. Xanterra narrativizes the dining experience at these restaurants in ways that align with the audience of conscious enrichment-seekers they imagine.

Also situated at Canyon Village, the Canyon Lodge M66 Grill provides one of the most immediate contrasts to the Eatery. The restaurant is temporarily closed as of summer 2022, as labor shortages and supply chain problems have forced concessioners nationwide to reassess capacity. But it will surely reopen, and, prior to COVID, it was one of the pricier reservations-only restaurants in the park. The lone picture on its website suggests a less whimsical or striking design aesthetic than one can find over at the Eatery: Mid Century modern chairs with colorful vinyl seat cushions fill an empty restaurant, while a bar in the background overlooks a window facing some evergreens. Lacking much in the way of visual excitement, Xanterra nonetheless ensures visitors that eating at the M66 Grill is an exclusive experience by communicating the challenging parameters for securing a table: "Dinner reservations can be made beginning May 1 for the following year with Canyon Lodge room reservations and 60 days in advance without Canyon Lodge room reservations." Travelers escaping the coastal cities for a vacation at Yellowstone will recognize from this reservation policy that a table here is something to be planned for, to be anticipated. There is a certain allure to getting a reservation that requires such planning—it is exclusive and requires the reserver to have done their homework to have insider knowledge.

Indeed, despite the casual dress code, the opportunity to experience "full-service dining in a casual Mission 66-themed setting" with "a menu of house-made soups, entrée salads, and creatively prepared beef, chicken, pork, Red trout and vegetarian dishes" (Xanterra 2019b) that is promised on the website is a harder get than some of the 50 best restaurants in the world. The 2019 dinner menu contains a prominent note in an eye-catching brown box, addressed to discerning diners:

At M66 Bar and Grill, we understand that our wonderful natural setting is part of a complex and fragile global ecosystem and must be protected and preserved for future generations. Therefore, we promote a sustainable model of fresh forward dining. We're softening our environmental footprint while bringing our guests the highest quality food and beverages.

(Xanterra 2019b)

In contrast to the paucity of local and sustainable mentions on the Eatery menus, there is no shortage of eco-sensitive good taste-conjuring on the M66 menu. The wines and beers on tap are all local, and the beer list mentions the whereabouts of each brewery (some from Bozeman, Missoula, Belgrade, and Red Lodge, MT, as well as one from Victor, ID). Almost all of the entrees get the vaunted green leaf: From the Red Bird Farms Natural Airline Chicken Breast (antibiotic and cage free, though not mentioned on the menu) to the Sauteed Red Trout (crusted with Montana Morado maize), the Bison Burger or Mixed Game Bratwurst, and several offerings from Case Custom Meats out of Yoder, WY, there is an abundant supply of place name-checking on the M66 menu. Among "shareables," there is a Montana cheese plate with Tucker Family Farms Feta, Amaltheia Dairy chevre, and Mountain Mocha alpine cheese, and there are also Lamb Sliders made with Montana Natural lamb. For dessert there's even a House-Made Flathead Cherry Crisp featuring "juicy Flathead cherries from Northern Montana under a crunchy, buttery oat crust." As expected, all these qualifiers come with a price tag, as entrees ranged from $15.75 to $28.95 on the 2019 menu. But these prices are still quite reasonable for the targeted audience. Not only do they allow the concessioner to meet market demand and turn a healthy profit, they also effect compliance with NPS sustainability guidelines (Aubrey 2013).

The M66 breakfast menu, while offering fewer options with the green leaf that signifies sustainable or organic, has a section explaining more about Xanterra's Softest Footprint goals: "We do it by finding products, where possible, that are: locally produced, organic, third-party certified and support sustainable farming, fishing and business practices" (Xanterra 2019a). They then list 32 partners, from Wheat Montana to Fat Robin Orchard.

Higher-End Eating: Other Yellowstone Options

The way Xanterra narrativizes on its website and menus for its higher-end park concessions is not unique to the M66 Grill, but slight variations on their key themes can be found. The Grant Village Dining Room, overlooking the scenic West Thumb geyser basin and Yellowstone Lake, plays equally hard to get with its reservation policy. But on its menu, the green leaf icon is not just for items made with sustainable or organic ingredients but also for those made within 500 miles—one of the only clear statements about foodshed seen in the dozens of Xanterra concessions menus that I reviewed. There is, however, less info provided about the sourcing of menu items, but there is a brief italicized statement, "*We proudly support local ranchers.*"

At the elegant, stately, white tablecloth Lake Hotel Dining Room, which overlooks the waters of Lake Yellowstone, the website tells us that "the menu is creative and upscale, with unique dishes of fresh fish, wild game and more. Our commitment to sustainable cuisine (local and/or organic) is no more prevalent than at this casually elegant restaurant." On this dinner menu, many of the green leaf items indicate a city, although it's not always clear what hails from there. The "Wyoming

Legacy Steak" is clearly from Cody, as stated, but what part of the "Caramelized Onion and Gruyere Cheese Ravioli" hails from Denver?

With a casual upscale (for the parks) vibe akin to the M66 Grill, the Mammoth Hotel Dining Room (MHDR) is a no-reservations venue known for its Art Deco style with views of the old Fort Yellowstone parade grounds and grazing bison and elk. Xanterra's website and the restaurant's menu lean hard into the notions of sustainability that entice their target market of travelers who fancy themselves conscious, responsible, and traveling lightly. The site bills the MHDR as "The First 4-Star Certified Green Restaurant in the National Parks," touting its 2011 certification by the Green Restaurant Association, a nonprofit that evaluates restaurants on environmental sustainability criteria including energy, food, water, waste, disposables, chemicals, pollution reduction, furnishings, and building materials.

> To achieve its 4-Star certification, the Dining Room demonstrated sustainable operations through a major restroom remodel, installation of energy-saving LED lamps and water-saving fixtures, sourcing of local and organic cuisine, recycling and composting restaurant waste, and using environmentally-preferable cleaning products.
>
> *(Xanterra 2022b)*

The MHDR dinner menu contains a handful of green leaves and concludes with a list of Xanterra's 26 local food-sourcing partners, as well as their Softest Footprint philosophy statement (Xanterra 2021). Although a restroom remodel alone might not be enough to attract a clientele of cosmopolitan diners, Xanterra's storytelling on its restaurant websites and menus works to affirm travelers' imagined selves. But perhaps no Yellowstone concession speaks to cosmopolitan travelers with a taste for enrichment as much as the Old West Dinner Cookout.

"An Evening You Won't Soon Forget": Old West Dinner Cookout

The range of concessions that Xanterra offers at Yellowstone National Park is typical of park concessions nationwide, particularly in larger, remote, yet heavily traveled parks. What is truly atypical in the parks is the dining experience as an attraction in its own right, as exemplified by the Roosevelt Old West Dinner Cookout. The boulder-strewn plains of Pleasant Valley, near Lost Creek, in the north central part of the park, are the site of one of the most unusual food experiences in a US national park: a pricey, alfresco, picnic-table or stump-based all-you-can-eat dinner consisting of a massive steak, coleslaw, potato salad, baked beans, corn, muffins, watermelon, fruit crisp, drinks, and cowboy coffee, only accessible via horseback or horse-drawn wagon.

To partake of the dinner, visitors plunk down $64–$99 per adult, depending on their choice of horse or wagon, and meet up at the Roosevelt Corral, where they are greeted by jeans-, plaid-, and cowboy hat-clad wranglers who lead them on horseback or drive them in a covered wagon. Either way, they end up at Yancey's Hole, a clearing in the Pleasant Valley, where they can listen to cowboy singers and storytellers, pet the horses, and scan the horizon for bison. They might notice that the cooks are busy preparing the dinner over propane-fueled grills in permanent, open-sided, industrial kitchens, rather than over a genuine provisional campfire, but they probably won't be too bothered by this. Once the dinner bell rings, it is every man for himself, and everyone gets in line for dinner. Travelers pass through a buffet line where "cowboys" pile food on their metal plates before they retire to an upturned stump or picnic table to chow down. Anyone still hungry can pass back through the food line, but most travelers gather around the fire to listen to more cowboy stories and to join in a sing-along before they're loaded back on their horses or wagons for the ride back to the corral.

Xanterra's old west cookout relies on different representational strategies than its basic or higher-end food concessions. They make few claims about the quality of the food, beyond describing it as "real cowboy grub" with "real western beef steaks" and "signature Roosevelt Baked Beans" (Xanterra 2022c). (Apparently, in this simulacrum of an old west dinner, that the food is real should not be taken for granted!) There is no information provided—either on the website or during the experience itself—about the sourcing of the food or its sustainability. With no photographs on the web page illustrating the food or the eating experience, the emphasis is on the mode of transport to get to the meal (horse or wagon), the beautiful outdoor setting, the massive servings, the ability to observe and interact with "cow folk" while eating, and the winkingly transformative nature of the evening that will take visitors from city slicker to cowpoke. Xanterra relies on playful prose full of dropped g's to emulate western cowboy talk:

> You'll find those cooks dishin' up some real cowboy grub at our popular Old West Dinner Cookout. The coffee's brewin' over the open campfire, and our wranglers love talkin' your ears off over a strong cup o' Joe!' … You'll find your boots tappin' to old western songs sung by our singin' cowboy. You may have come here as a city slicker, but you'll go back as a regular cowpoke!
>
> *(Xanterra 2022c)*

The transformative promise of the old west cookout is contingent on the acceptance of its simulated authenticity. Xanterra beckons visitors to become enriched through a one-of-a-kind (thus genuine) food experience in which the food itself—whether it is local, organic, sustainable, or any of the other things touted in the representation of the regular concessions—is almost an afterthought, taking back seat to the intimate, spectacular, immersive, and ultimately unforgettable nature of the event.

Non-Restaurant-Specific Representation: Xanterra
Special Features

Beyond the way Xanterra represents park foodways on its restaurant-specific web pages, on concession menus, and at concessions events like the cowboy cookout, the company also communicates important values to consumers through a cadre of feature stories that rotate through its website. The easiest place to find these features is in the "Read More About Our Dining" section at the bottom of any of the restaurant-specific pages; links to specialized stories that reflect well on Xanterra abound. These stories center on two themes: sustainability and diversity of offerings.

The most noticeable theme in Xanterra's web-based special features is sustainability. A 2018 interview with Yellowstone Executive Chef Mike Dean connects Xanterra's local sourcing practices with the founding logic of the park:

> The idea of using and providing locally produced goods has become something of a culture here at Yellowstone National Park Lodges. The whole idea of nature, sustainability and conservation started with the first National Park, so it only makes sense to continue and enhance that practice. We try to source as many products as we are able locally, and work with growers, ranchers, and distribution companies that have the same practices in mind.
>
> *(Dean, cited in Xanterra 2018b)*

Dean acknowledges that small local producers in Wyoming and Montana can't necessarily provide enough products to supply Xanterra's Yellowstone NP concessions through a whole season but says that he finds "niches" to showcase the products as they are available. Roosevelt Lodge, for instance, offers a shepherd's pie made with beef from cattle grown within 40 miles of the kitchen, but the beef is used for that one menu item only (Xanterra 2018b). Another Yellowstone Hot Spot, this one in the form of a "Top Ten Things to Eat and Drink in Yellowstone" infographic, doesn't talk about the restaurants or concessions directly, but also strikes the theme of the delicious local food and drink available: Huckleberries, trout, bison, anything made in Montana, and local beers are particularly touted (Xanterra 2018a). And sometimes the stories zoom in on the local small producers that provide food to Xanterra in Yellowstone: a profile of Belgrade, Montana-based Amaltheia Organic Dairy owner Sue Brown kvells about how Xanterra's willingness to feature her goat cheese, organic beets, and baby kale on the Lake Hotel Dining Room's menu has boosted her sales and sent a lot of new customers her way (Nugent 2017).

Other articles concentrate on sustainability, emphasizing Xanterra's commitment to sustainably produced food and beverage purchases and sustainable management practices. A 2017 article on the website emphasizes the food waste reductions at its food concessions in the parks, as well as innovations like keg beer, aluminum water cans, and biodegradable cutlery that have allowed Xanterra to minimize single-use plastics and remove bottles from the waste stream (Xanterra 2017b). Another 2017 article applauds the company for reaching the 50% mark for sustainable food and beverage purchases in its Yellowstone food concessions. The article goes on to acknowledge that "the goal is to increase them to 70 percent by 2025" and implores visitors to do their part by ordering local and sustainable menu items and applying the lessons learned in Yellowstone in their home communities (Loomis 2017). Another piece that rotates frequently through the Xanterra site trumpets the extent to which sustainability is an essential part of the company culture. It explains how consumer glass waste from the food service operations in the park has been recycled into park benches, picnic tables, and counters, and the author exhorts readers to do their part:

> So the next time you enjoy some locally brewed beer or Montana syrup during your stay in Yellowstone, be sure to throw the empty bottle into one of the many recycling bins throughout the park. It just might turn up in the next round of green furniture or building elements added to the park.
>
> *(Stoddart 2016)*

Xanterra's storytelling around sustainability hails the conscious traveler, allowing them to imagine themselves as playing an important role in conservation and preservation through their consumption choices (buying the small-producer goat cheese and looking out for the local beef in the Roosevelt shepherd's pie) and citizen practices (recycling and patronizing companies like Xanterra with a commitment to sustainability).

The other prominent theme in Xanterra's non-restaurant-specific stories is its diversity of offerings. In a Yellowstone Hot Spot interview that circulates widely on Xanterra's web pages, Yellowstone Executive Chef Mike Dean argues that one of the company's greatest strengths is its variety, from fine dining to casual dining in both historic and new buildings: "I appreciate that we try to find a balance of food styles to accommodate all guests, appealing to visitors from around the world" (Dean, cited in Xanterra 2018b). This kind of narrative often illustrates the volume of Xanterra's service, suggesting that the food concessions offer something for everyone: "Last year, the park's 4 million-plus visitors consumed 232,120 hamburgers, 69,580 pounds of scrambled eggs and 56,209 gallons of ice cream, among other fare" (Clark 2018).

As much as the immediate message seems to be about Xanterra's democratic "something for everyone" impulses, there exists a less obvious implication about class and distinction entwined with this theme. Usually, when the "diversity of options" theme shows up in Xanterra messaging, the company emphasizes how their range and scale allows them to cater to those with more discerning palates. For instance, in the very same story that describes the 56,029 gallons of ice cream consumed in 2017—a sure feat of volume for the satisfied masses—Executive Chef Dean notes that huckleberry ice cream is the best-selling flavor in the park, again speaking to its mass appeal (Clark 2018). This does not, at first, read like messaging that is aimed at sophisticated travelers, but the hook is in the details. The article mentions that the popular huckleberry ice cream is made by a local Montana business, reminding the target "conscious traveler" of the virtuous aspect of consuming the product—supporting small businesses and reducing the ecological consequences of long-haul food transportation. The story goes on:

> In fact, any dish incorporating huckleberries is a guaranteed winner, Dean notes. The impossible-to-cultivate berry grows wild in the region and must be harvested by hand. You can savor its sweet/tart flavor in pancake syrup on breakfast tables throughout the park; in the smoked half chicken with huckleberry-chipotle barbecue sauce at Roosevelt Lodge; in the white chocolate and huckleberry cheesecake at Old Faithful Snow Lodge; and in the huckleberry crème brûlée in the Grant Village Dining Room.
>
> *(Clark 2018)*

The qualifiers of "impossible-to-cultivate" and "harvested by hand" and the range of dishes and eateries in which one can find the huckleberry suggest that a different kind of reader is being addressed—a Romantic one who admires the hard to get, the artisanal, the handmade and the non-mechanized—and who appreciates the opportunity to flex their knowledge and taste in making consumption choices throughout the park.

Another Yellowstone Hot Spot story from 2017, featuring an interview with Yellowstone NP Food and Beverage Director Lu Harlow, makes a similar move. Harlow first emphasizes volume: "We serve 50,000 pounds of prime rib in one year!" (Xanterra 2017a). But she goes on to explain how menu design in the park must consider the tradition and vibe of the facility "while reflecting the tastes of the clientele" (Harlow, quoted in Xanterra 2017a). Harlow explains that the menu at the iconic Old Faithful Inn Dining Room, with its popular dinner buffet, strip steak, and macaroni and cheese, reflects the broad demographic of the hotel, as well as its unique history as an iconic western lodge. In contrast, the recently renovated Lake Yellowstone Hotel resulted in "an enhancement of its upscale nature." Harlow said that this meant that the menu needed to be "upper end" to meet the changing guest requirements: "Creatively prepared fresh fish and locally produced beef tenderloin have been favorites. And the beef at Lake Hotel is very fresh, with field to table timing at about a month" (Harlow, quoted in in Xanterra 2017a). She relayed an anecdote, sharing that a wealthy long-time guest with homes in New York City and Jackson Hole (both high-end real estate markets) once called her directly to share his review of a meal at the Lake Yellowstone Hotel Dining Room. Harlow was nervous at first, then delighted to hear him say that "the meal and service was on par with the Big Apple. For Lu Harlow, there was no better compliment" (Xanterra 2017a). The story invokes concepts of creative preparation, local production, freshness, and successful catering to upscale taste to appeal to cosmopolitan travelers, who are assured that the things they value can be part of the dining experience in Yellowstone.

A final example of the "diversity of offerings" theme also showcases cosmopolitan values while describing the abundance of local produce and meats available in national park restaurants. In a rhetorical move reminiscent of the hipster line "I liked it before it was cool," a story about farm-to-table dining describes it as "undeniably one of the hottest culinary trends in the country right now" (Xanterra 2016), but turns an aw-shucks about-face in the next instant: "But America's national parks have been serving this style of cuisine for years, simply out of practicality" (Xanterra 2016). The practicality, it turns out, is the "geographic locations" of the parks that are "optimal for sourcing area specialties" (Xanterra 2016). The story gives a special shout-out to the Lake Yellowstone Hotel Dining Room, with "the largest sustainable selection in the park" and a "sophisticated setting." Overall, the narrative emphasizes seasonality, noting that visitors can "taste authentic local and regional flavors — especially at harvest time," in a move that speaks to cosmopolitan travelers who are suspicious of the industrial standardization that they might associate with park concessions.

Conservation vs Enjoyment? Swallowing Competing Mandates

I want to think about the eating experiences that Xanterra tells the world it offers in Yellowstone—from picnicking or grabbing a quick bite at Canyon Lodge Eatery to nailing a coveted reservation at Lake Yellowstone Hotel Dining Room or climbing on the saddle to head out to the Roosevelt Old West Dinner Cookout. To what extent do these experiences prioritize conservation? Or are they all about enjoyment? And do they allow for an experience of public lands characterized by using food as a response to social, economic, and political marginalization, the need for which was described in Chapter 3?

Despite being an industrial concessioner, as my close reading of their corporate narratives shows, Xanterra is no slouch in delivering fresh, local, sustainable food options, thus manifesting the conservation element of the national parks promise. However, not all food concession venues are made equal with regard to conservation. Dining spectacles like the Old West Cookout appear to be entirely independent of Xanterra's sustainability-related priorities. And basic cafeterias, at the level of the Canyon Lodge Eatery, may nod to concepts of sustainability and local sourcing but provide few options and very little information that allows eaters to "do the right thing" as they customize their meals. True conservation priorities are articulated and expressed most clearly in the park's higher-end dining experiences, where cosmopolitan diners are courted with detailed information about product sourcing and company commitments to the environment as they exercise their taste proclivities.

In earlier chapters, I outlined the kind of traveler the national parks were built for—those who were better off, seeking distance from the madding crowd, looking for an authentic experience with nature that would allow them to reflect and change their lives for the better. The current dining experiences available at national parks to a large extent miss their mark with the cosmopolitan descendants of this crowd. Even with enhanced efforts around local food sourcing and sustainable practices, something is missing. At the lower end, there's not much information about food and conservation, or there aren't many options. At the higher end, information abounds, and the reservation policies satisfy the desire for exclusivity. But in many instances, like in Canyon Lodge M66 Grill, there is no communing with nature to be done while dining. Almost theatrical experiences like the Roosevelt Old West Cookout provide kitschy cachet in today's experience economy and capitalize on a truly spectacular natural environment, but provide little at all in the way of a conservation-minded dining experience.

In conclusion, a glimpse of the way that food concessions are narrativized at Yellowstone demonstrates how the kinds of tensions between conservation and enjoyment that have been inherent in the parks since they were founded play themselves out. The US relies on industrial food concessions in the parks, but the stories told by corporate giants like Xanterra encourage the visitors who are paying any attention—and the cosmopolitans certainly are—to swallow any unreconciled contradictions, safe in the knowledge that even if the most common ways of eating in the parks seem to undermine the founding premise of these lands, their individual acts of Romantic consumption support a better and more sustainable foodscape. Chapter 5 examines how narratives about aspirational dining experiences in the parks from concessioners, lifestyle media, and restaurant visitors themselves shape and reflect cosmopolitan priorities that set the stage for the kinds of bioregionalist and Indigenous-centered ecologies that are explored in more depth in Chapter 6.

Note

1 As of August 2023, there are no food concessions in 27 parks: American Samoa, Arches, Biscayne, Black Canyon of the Gunnison, Canyonlands, Capitol Reef, Congaree, Cuyahoga Valley, Gates of the Arctic, Great Sand Dunes, Guadalupe Mountains, Haleakala, Hot Springs, Indiana Dunes, Joshua Tree, Kenai Fjords, Kobuk Valley, Lake Clark, Mammoth Cave, New River Gorge, North Cascades, Pinnacles, Redwood, Saguaro, Theodore Roosevelt, Wind Cave, and Wrangell St. Elias (www.nps.gov/subjects/concessions/concessioners-search.htm#).

References

Aubrey, Allison. 2013. "Hold the Hot Dog: National Park Visitors Can Feast on Bison Burgers." *NPR.org*, June 7. https://www.npr.org/sections/thesalt/2013/06/07/189270752/hold-the-hot-dog-national-park-visitors-can-feast-on-bison-burgers.

Bartlett, Richard A. 1985. *Yellowstone: A Wilderness Besieged*. Tucson: University of Arizona Press.

Blevins, Jason. 2013. "Xanterra Parks and Resorts Is Expanding Beyond National Parks Concessions." *The Denver Post*, July 26. https://www.denverpost.com/2013/07/26/xanterra-parks-resorts-is-expanding-beyond-national-park-concessions/.

Boss, Donna. 2019. "A Retro-Renovation for the Ages in Yellowstone National Park." *Food Service Equipment and Supplies*, May 1. https://fesmag.com/topics/project-profiles/facility-design-project-of-the-month/16806-a-retro-renovation-for-the-ages-at-in-yellowstone-national-park.

Clark, Jayne. 2018. "Eat Up! What's on the Menu in Yellowstone." Posted October 3, 2018. https://www.yellowstonenationalparklodges.com/connect/yellowstone-hot-spot/eat-up-whats-on-the-menu-in-yellowstone/.

Concerned Citizen. 2018. "Stop Starbucks in Yosemite." Accessed August 10, 2022. https://www.change.org/p/stop-starbucks-in-yosemite.

George, Richard. 2021. *Marketing Tourism and Hospitality*. London: Palgrave Macmillan.

Government Food Service. 2017. "National Park Service." Published October 2017. https://ebmpubs.com/GFS/GFSdata/2018_GFS_Almanac/11_NationalPark Service.pdf.

Hartzog, George B., Jr. 1988. *Battling for the National Parks*. Mt. Kisco, NY: Moyer Bell Limited.

Herman, Dennis J. 1992. "Loving Them to Death: Legal Controls on the Type and Scale of Development in the National Parks." *Stanford Environmental Law Journal* 11 (3): 3–67.

Loomis, Christine. 2017. "Green Dining in Yellowstone." Posted March 8, 2017. https://www.yellowstonenationalparklodges.com/connect/yellowstone-hot-spot/green-dining-in-yellowstone/.

Mantell, Michael. 1979. "Preservations and Use: Concessions in the National Parks." *Ecology Law Quarterly* 8 (1): 1–54.

Meeker, Joseph W. 1973. "Red, White, and Black in the National Parks." *The North American Review* 258 (3): 3–7.

Miller, Max. 2017. "'Yellowstone Is Its Own Giant' When It Comes to Worth of Private Businesses." *CodyEnterprise.com*, October 9. https://www.codyenterprise.com/news/local/article_8346e6b2-ad2e-11e7-9cee-4f8da954f461.html.

National Park Service. 2006. "Management Policies." Accessed August 10, 2022. https://www.nps.gov/subjects/policy/upload/MP_2006.pdf.

National Park Service. 2021. "2021 National Park Visitor Spending Effects Economic Contributions to Local Communities, States, and the Nation." Natural Resource Report NPS/NRSS/EQD/NRR—2022/2395. Accessed August 11, 2022. https://www.nps.gov/subjects/socialscience/vse.htm.

National Park Service. 2022a. "Authorized Concessioners." Last updated June 23, 2022. https://www.nps.gov/subjects/concessions/concessioners-search.htm.

National Park Service. 2022b. "National Park System." Accessed August 10, 2022. https://www.nps.gov/aboutus/national-park-system.htm.

National Park Service. 2022c. "Visitation Numbers." Accessed August 8, 2022. https://www.nps.gov/aboutus/visitation-numbers.htm.

Nugent, Karley. 2017. "Featured Farm: Amaltheia Organic Dairy." Posted September 7, 2017. https://www.yellowstonenationalparklodges.com/connect/yellowstone-hot-spot/amaltheia-farm-qa/.

Peterman, Audrey and Frank Peterman. 2009. *Legacy on the Land: A Black Couple Discovers our National Inheritance and Tells Why Every American Should Care*. Plantation, FL: Earthwise.

Stoddart, Veronica. 2016. "Giving Beer Bottles New Life." Yellowstone National Park Lodges. Posted December 2, 2016. https://www.yellowstonenationalparklodges.com/connect/yellowstone-hot-spot/giving-beer-bottles-new-life/.

Taylor, Dorceta E. 2016. *The Rise of the American Conservation Movement: Power, Privilege, and Environmental Protection*. Durham, NC: Duke University Press.

US Department of the Interior. 2017. "Hunting and Fishing on National Parks and Fish and Wildlife Refuges." Posted March 1, 2017. https://www.doi.gov/blog/hunting-and-fishing-national-parks-and-fish-and-wildlife-refuges.

Weber, Joe and Selima Sultana. 2013. "Why Do So Few Minority People Visit National Parks? Visitation and the Accessibility of 'America's Best Idea.'" *Annals of the Association of American Geographers* 103 (3): 437–64.

Weissman, Arnie. n.d. "The Enigmatic Travel Giant. Xanterra Travel Collection, Explored and Explained." *Travel Weekly*. Accessed August 11, 2022. https://www.travelweekly.com/Travel-News/Tour-Operators/Xanterra-enigmatic-travel-giant.

Xanterra. 2016. "Tasting Nature: Farm-To-Table Dining at the National Parks." Posted December 1, 2016. https://www.xanterra.com/stories/culture-lifestyle/tasting-nature-farm-to-table-dining-at-the-national-parks/?_ga=2.243646529.578633855.1583590343-1399565860.1583590343.

Xanterra. 2017a. "Serving Good Food and Supporting Local Communities." Yellowstone Hot Spot. Posted December 12, 2017. https://www.yellowstonenationalparklodges.com/connect/yellowstone-hot-spot/serving-good-food-and-supporting-local-communities/.

Xanterra. 2017b. "Turning Food into Flowers and Cooking Oil into Fuel. How Xanterra-Managed National Park Lodges are Reducing Food Waste and Supporting Sustainability." Posted on September 8, 2017. https://www.xanterra.com/stories/parks/turning-food-into-flowers-and-cooking-oil-into-fuel/?_ga=2.184010594.578633855.1583590343-1399565860.1583590343.

Xanterra. 2018a. "Yellowstone Hot Spot, Infographic: Top Ten Things to Eat and Drink in Yellowstone." Posted April 26, 2018. https://www.yellowstonenationalparklodges.com/connect/yellowstone-hot-spot/infographic-top-10-things-to-eat-and-drink-in-yellowstone/.

Xanterra. 2018b. "Yellowstone Hot Spot, Q&A with Yellowstone Executive Chef Mike Dean." Posted March 20, 2018. https://www.yellowstonenationalparklodges.com/connect/yellowstone-hot-spot/q&a-with-yellowstone-executive-chef-mike-dean/

Xanterra. 2019a. "M66 Breakfast Menu." Accessed September 15, 2019. http://www.yellowstonenationalparklodges.com/content/uploads/2019/05/CL-M66-Breakfast-S19.pdf.

Xanterra. 2019b. "M66 Dinner Menu." Accessed September 15, 2019. https://web.archive.org/web/20211103125333/http://www.yellowstonenationalparklodges.com/content/uploads/2019/05/CL-Dinner-S19.pdf.

Xanterra. 2021. "Mammoth Hotel Dining Room Menu." Accessed August 12, 2022. http://www.yellowstonenationalparklodges.com/content/uploads/2017/04/MHS-Dinner-S21.pdf.

Xanterra. 2022a. "Give the Gift of Travel." Accessed August 11, 2022. https://www.xanterra.com.

Xanterra. 2022b. "Mammoth Hotel Dining Room." Accessed August 12, 2022. https://www. yellowstonenationalparklodges.com/restaurant/mammoth-hotel-dining-room/.

Xanterra. 2022c. "Old West Dinner Cookout." Accessed August 12, 2022. https://www. yellowstonenationalparklodges.com/adventure/wild-west-adventures/old-west-dinner-cookout/.

Xanterra. 2022d. "Our Fred Harvey Legacy." Accessed August 11, 2022. https://www. xanterra.com/who-we-are/our-fred-harvey-legacy/.

5

REPRESENTING UPSCALE RESTAURANTS

US national parks were born of Romantic notions that glorified nature (in opposition to human-made industry) and emphasized the importance of individual experiences of solitude for spiritual renewal. Their founding was urged by private interests with profit motives in providing railroad travel, hotels, and concessions to park visitors. Most food concessions in today's parks are managed by large-scale operations: International, national, and regional hospitality corporations dominate the landscape. In this respect, the profit motives of the private interests have prevailed, and appear to be in direct tension with the anti-industrial, preservation-minded, self-making impulses of the Romantics. While Chapter 4 explored how the narrative rhetoric of the concessioners helps travelers to swallow these tensions, this chapter considers contemporary representations of eating in upscale US national park restaurants as sites for the negotiation of status and explores how digital media are both reflecting and shaping discourses of health, sustainability, the local, landscape, and cosmopolitanism today.

Fine Dining in the National Parks

Many US national parks feature a range of dining options, offering everything from do-it-yourself picnic areas to camp stores, grills, coffee shops, cafeterias, and even relatively fancy restaurants. I have selected the fancier restaurants as sites for analysis here, but not necessarily because they are profit centers; it is often the case that full-service restaurants, with their higher food, facilities, and labor costs, operate on a slimmer profit margin (3%–5% on average) than fast-food restaurants or food trucks with lower overhead (6%–9% on average) (Sheetz 2020). Rather, the full-service restaurants are most likely to be housed in historic structures in the most iconic (and heavily trafficked) settings of the parks and to sell themselves as offering

DOI: 10.4324/9781003455516-6

a dining experience that has symbolic value. Thus, their high visibility allows concessioners a brightly lit stage on which to display their ideological commitments.

Representationally speaking, the National Park Service (NPS) doesn't exactly go overboard to shill for park concessioners on its websites. Most US parks' websites have a top-level navigation that includes "Plan Your Visit," then a sub-navigation menu for "Eating and Sleeping," but what level of detail one finds there depends on the park. Many park sites emphasize picnicking, like Acadia, which has a page devoted to "Eating and Picnicking" (National Park Service 2021) in which most of the space is dedicated to information about picnicking. After a long listing of picnic sites, a map, and picnicking rules and regulations, almost as afterthought, one finds the subheading "Restaurants" at the very bottom of the page. This information hierarchy is suggestive: The restaurants in the parks are clearly not the main attraction from the perspective of the NPS, but the restaurants themselves and their corporate managers deploy more aggressive representational strategies elsewhere to cement their brands.

A word about fancy, park-style: What passes for fine dining in the US national parks is a far cry from the formal service, refined décor, and expensive, top-quality food of the traditional fine-dining sector. While the restaurants are full service, and reservations are important for some of them, dress codes are virtually nonexistent, service is relatively informal, and the quality of food is generally not on par with even mid-level restaurants outside park borders. The view, though, is often priceless.

I have chosen to look closely at three of the most high-end restaurants in US national parks—Jordan Pond House in Maine's Acadia National Park, Metate Room in Colorado's Mesa Verde National Park, and Jenny Lake Lodge Dining Room (JLLDR) in Grand Teton National Park in Wyoming, all run by different concessioners. They are among the restaurants frequently cited as offering the best dining in the national parks (Sulem 2016). Although they share the distinction of offering some of the finest meals in the parks, their public profiles and reception by eaters vary in interesting ways.

Key Themes Reflecting Cosmopolitan Taste: Health, Sustainability, Landscape, and Local/Indigenous Culture

Across these three examples, in the way the restaurants are represented (primarily via their websites and menus) by the companies that manage them, professional media representations in the form of reviews and other web content covering them, and in the responses of restaurant visitors on social media (Instagram) and review sites, several themes emerge that reflect the priorities of cosmopolitan taste: health, sustainability, landscape, and local/Indigenous culture. This chapter examines these discourses in representations of the in-park fine-dining experience at these restaurants.

First, health: Although it may seem like a bonus for healthy options to be available at these restaurants, it is in fact mandated by the NPS and is a condition of the concession contract. All concessioners with contracts effective in 2016 or later must meet standards pertaining to food ingredients and choices, food preparation, and visitor/consumer education in any front-country operations (National Park Service 2017). The requirements involve things like integrating fruit, vegetables, and whole grains into the menu; reducing fat, sodium, and added sugar content in some offerings; offering lite dishes and smaller portion options; and educating consumers about their healthy choices. Despite these requirements, discourses of health are rather underplayed by the fine-dining establishments themselves, the consumers who eat at them, and the professional media that cover them.

Second, sustainability: Sustainable food options are not mandated by the NPS, which recognizes that they "can be more difficult to obtain and more expensive," although they are encouraged (National Park Service 2017). Specifically, the NPS suggests offering sustainable seafood options; fair trade and shade-grown coffee; and products and ingredients that are seasonal, local, or regionally produced (in-state or less than 400 miles away), organic, or free from hormones or antibiotics. The NPS also encourages educating consumers about sustainable options. As I discussed in Chapter 4, higher-end restaurants are more likely to showcase sustainable food options than their less formal counterparts. The more detailed analyses that follow in this chapter suggest that sustainability is an important theme used by high-end park restaurants when they represent themselves, and that is also important in the discourse of professional media and everyday consumers as well. Associating local products with economic and social sustainability attracts visitors who "cast themselves into the role of the 'good' and 'responsible' tourists who care about the destinations they are visiting" (Sims 2009, 334). This all sounds very skeptical, and the neediness of it all may certainly prompt an eyeroll, but economically and socially sustainable behavior is a good thing, even if already privileged people get yet another halo for supporting it.

Third, landscape: Like the parks themselves, many of the more upscale restaurants inside the parks are known for their views of the sublime. In older parks, with structures built before the mid-twentieth century, the most classic lodges tended to be sited to maximize visitors' views. From the steps of lodges like Old Faithful Inn in Yellowstone National Park, El Tovar in Grand Canyon National Park, Ahwahnee in Yosemite National Park, and Crater Lake Lodge, one can take in the landscape that has occasioned travel to the park. In fact, many of the old dining rooms in these lodges were built with windows facing the main attraction, whether geyser, canyon, mountains, or lake, to the delight of their diners. Mission 66, introduced in 1956, zoned restaurants and other hubs for human activity away from spaces of natural beauty (Clayton 2019). Nonetheless, landscape and views continue to register as chief selling points of upscale park restaurants and top interests of would-be diners.

Fourth, local/Indigenous culture: Another theme in the representation of upscale national park restaurants is the incorporation of Indigenous or local foods or other cultural elements. The very globalization that allows national or multinational food conglomerates to win contracts for food concessions in the national parks may be perceived as a considerable threat to local gastronomic identities; thus, the large concessioners are spurred to demonstrate that they are creating new opportunities to showcase and reinvent local gastronomic products and identities (Mak, Lumbers, and Eves 2012) as a way of mitigating their perceived negative impact. Local food sourcing is often understood as a pillar of sustainable foodways but is also popular because local foods are imagined as authentic symbols of place, thus reinforcing tourist beliefs that they are experiencing the genuine culture of the place they are visiting (Sims 2009). Indicating provenance, or *terroir*, is the most common discursive strategy to suggest that food is worthwhile for foodies (Johnston and Baumann 2007, 179). The analysis in this chapter indicates that local food sourcing is a prominent theme in the way corporate managers represent national park restaurants and is an interest among many of those who eat there. A related but less dominant theme, that nevertheless bears examination, is the incorporation of Indigenous culture or foodways into the upscale park dining experience; as the following analysis indicates, Indigenous culture is more likely to be integrated into the décor than the menu, although there are some breakthrough exceptions to this rule.

These themes combine to indicate that upscale dining experiences in national parks are suited to cosmopolitan tastes. Representations promise that upscale venues in the parks provide high-quality food and a memorable experience for discerning diners.

> As control over the food supply is increasingly consolidated by transnational agribusiness, encounters with complex, flavorful food will even become museum-like and rare. Food tourists of the future will seek out the rare food places of the world. [...] The food tourist is increasingly challenged to obtain this experience within a food system of increasing uniformity and blandness. [...] There is no adventure in the taste of industrially produced fare.
>
> *(Hansen 2015, 51)*

Given the widespread reputation of national parks as food deserts, corporate concessioners and eaters alike are deeply invested in foodie discourses (Johnston and Baumann 2015) that tell a different story about the distinctive possibilities available.

Modes of Representation: Restaurant Websites, Menus, Professional Coverage, Review Sites, and Instagram

To get a clear picture of how upscale park restaurants are represented, it is necessary to look from a variety of angles. Concessioners represent restaurants through both their websites and their menus. Sometimes, gatekeepers—whether newspaper columnists, guidebook authors, or even food bloggers—will contribute to the representational enterprise. On the receiving end are the eaters, who have become anything but passive consumers. Global flows of media seed cosmopolitanism, as they allow audiences to position their lives relative to global others (Appadurai 1990). This encourages audience participation in global consumer culture. Eaters increasingly represent restaurants on review sites like Tripadvisor and Yelp as well as through social media, like Instagram, and their reviews reflect and shape the discourses I have outlined. National parks increasingly provide visitors with encounters across cultural differences and opportunities to perform cultural competence, particularly regarding food (Figueiredo, Pico Larsen, and Bean 2021). In studying the way that these experiences are narrated by eaters on review sites, I'm mindful of Herman's investigation of review sites to determine the ways restaurants engage cosmopolitan sensibilities by using language, imagery, and interpretation clues (Herman 2021, 5). Of course, reviews are a not an exhaustive index of responses to a dining experience, but they do constitute an important and influential discursive form.

Websites and menus are the most obvious means for restaurants to represent their identity and commitments to consumers. Jurafsky (2014) contends that menus offer clues about how people think about wealth, status, and food, and that upscale restaurants tend to offer fewer choices than are available at casual eateries and to use few words in general on their menus, but those words are often long and from high-status foreign languages like French. Most restaurants today use their web pages not only to post menus but to extend this framing as part of marketing efforts to impact consumer behavior (Litvin, Blose, and Laird 2005). Restaurant websites are one of the most important information sources for consumers, but for upscale restaurants, they are considered alongside impressions of the restaurant's reputation, ratings by gatekeepers, and recent reviews in determining where to eat (Yilmaz and Gültekïn 2016).

In the age of mass media, professional food critics—experts with formal training and exclusive experience—were the dominant gatekeepers for knowledge about restaurants. Their reviews in print media sources like newspaper columns, guidebooks, and lifestyle magazines clued readers in to dining experiences deemed worthwhile, relevant, and high status (Johnston and Baumann 2015; Feldman 2021). Many of these media sources have become digitized, and today's internet offers a bevy of gatekeeper-crafted restaurant coverage. Digitalization has led to a moment where the gatekeeping function of expert cultural intermediaries is eclipsed by the efforts of amateurs, everyday people who register their judgments about food and restaurants in digital spaces—including review sites and social media—with great regularity (Jurafsky et al. 2014; Lupton 2016; Zeamer 2018; Kobez 2020; Lewis 2020; Contois and Kish 2022). Indeed, digital culture "provides the contemporary *infrastructure* and *grammar* through which food is increasingly communicated and fought over. Both are mundane technologies of identity making. Both have material dimensions. And both hold structural, symbolic and ideological significance"

(Feldman and Goodman 2021, 1227). When they were first created in the 1870s, US national parks were technologies for reproducing nature using the aesthetic forms of landscape representation available then; today, it makes perfect sense that the parks, and the restaurants they house, will remediate nature using the aesthetic forms of digital media (Grusin 2004, 172).

Among amateur cultural intermediaries, the group known as cosmopolitans are particularly influential, especially when mediating spaces like national parks. Bell and Hollows (2007) note that cosmopolitan subjects are allowed choice, mobility, and change, while those who represent the local are static keepers of the "authentic" ways with which cosmopolitans commune. Cosmopolitan travelers, who are contemporary, adventurous, omnivorous, and multicultural, are produced in a culinary tourism imaginary that frames them as distinct from locals, who are thought of as monocultural, exotic, and old-fashioned (Leer 2019). Cosmopolitans, then, thrive on the presence of locals (especially Indigenous people and their foodways) in national parks; as Hannerz writes, "there can be no cosmopolitans without locals" (Hannerz 1996, 11). In Chapter 6, I look more closely at how this particular (and problematic) cosmopolitan desire has surprisingly transformative benefits for recentering Indigenous foodways in the parks; in the meantime, given cosmopolitan concerns with authenticity and open-mindedness, it's especially instructive to look at how cosmopolitan interests play out in the spaces between professional gatekeeper-produced restaurant coverage and reviews produced by amateurs on review sites and in social media. Careful examination of the food discourses in these spaces—how they manifest familiar privileges and anxieties over status (Mapes 2018) signaled by "the classed consumption of racialized products" (Gualtieri 2021, 906)—reveals the ideological underpinnings of social class and ethnoracial formations in a manner that updates the connection between national parks and aesthetic forms.

Jordan Pond House

In the northeastern US, in the state of Maine, Acadia National Park's Jordan Pond House has existed in some iteration since 1893. Known for its sprawling lawn overlooking both scenic Jordan Pond and the iconic twin mountains known as The Bubbles, as much as for its tradition of popovers and tea, Jordan Pond House is Acadia's only dining option beyond picnicking. The concession is operated by Dawnland LLC, whose parent company Ortega has concession contracts in at least seven small-to-midsize national parks and many state parks. Like Xanterra in Chapter 4, Ortega/Dawnland promotes popular discourses of sustainability as they emphasize tradition, conservation, and landscape. However, except for landscape, these interests for the most part do not show up in eater representations.

Apart from government regulations and guidelines, one would think that any fine-dining restaurant in a national park would be wise to emphasize sustainability anyhow, as this is a key food discourse for the cosmopolitan visitors who are the target audience and who have the most power to circulate information about the restaurants (Wise and Velayutham 2009). The website of Jordan Pond House highlights the restaurant's healthy, local, and sustainable food, and provides navigation to a page on environmental stewardship, including a downloadable "environmental policy." The Jordan Pond House website mentions Dawnland LLC's commitment to water preservation, energy reduction, waste management, sustainable purchasing, and preservation and protection of park resources.

Nolan Capital, the private equity firm that took a majority interest in Ortega National Parks, LLC, in 2019, similarly touts the importance of their role as stewards:

> ONP has had a long history of service and it is our mission to continue as a responsible steward of our park partner's assets and resources. We are committed to serving our guests and visitors while respecting the environment and furthering the mission of these great public assets.
>
> *(Chairman Peter Nolan, quoted in Repanshek 2019)*

For its food recovery efforts in other parks, Ortega National Parks has been recognized by the Environmental Protection Agency. Ortega also touts its commitment to recycling and the use of recycled products.

Beyond its website, the most obvious way a restaurant represents itself is through its menu. The downloadable version of the Jordan Pond House menu provides valuable real estate to a statement titled "Fresh, Local, Sustainable":

> Ingredients retain more flavor and vitamins when they are fresh. Buying ingredients locally means they get to our kitchen and to your plate quickly for the freshest, tastiest meal possible. Local buying also reduces pollutants from long distance shipping. We are proud to include many organic ingredients and to support local farms and fisheries. This menu includes hundreds of sustainable, locally sourced, fresh ingredients.
>
> *(Dawnland LLC 2021b)*

Health claims do not get the same level of called-out treatment on the menu, perhaps because diners have reported finding health claims unappealing (Turnwald et al. 2017). However, the restaurant's website devotes a section to healthy eating, emphasizing the degree to which Jordan Pond House provides healthy choices. So, between the website and the menu, corporate managers are trying to make a clear statement about commitments to sustainability, which would seem to be of utmost importance when their food-service enterprises are situated within once-pristine landscapes dedicated to conservation.

The Jordan Pond House menu addresses the sophisticated type of eater who might be expected to visit the Park's only fine-dining establishment. The much-vaunted popovers, at $8 for 2 in 2021, are among the most affordable options, flanked by appetizers ranging from roughly $14 to $18, a $16.50 bowl of chowder

(!), and prix fixe dining at around $39–$45 per person—relatively high prices. As Jurafsky et al. (2016) have found, distinction is framed in menus by language connoting authenticity and abundance, which assumes educational capital on the part of the reader and yet does not reveal anxiety by oversignaling quality (Liberman 2004), for instance, through overzealous adjective use.

On the Jordan Pond House menu, authenticity is communicated via references to the provenance of the food (e.g., *local* lobster stew, *farmers market* vegan stew, *field* lettuces, *cage-free* boiled eggs, *heirloom* tomatoes, *Maine* wild blueberry sorbet, *Pineland Farms* smoked cheddar) and tradition (e.g., *Acadia Favorite* fresh seasonal catch, a Caesar salad with *classic preparation*). There are also plenty of fancy words that presume a reader with some degree of culinary capital—*gastrique* and *muddled*, for instance, are not entirely legible to the typical middle-class American who hasn't done a stint in culinary school. There is nothing much in the way of menu language signaling generous portions, but there are options aplenty in terms of the sheer number of choices: 29 food options and a startling 44 beverages, with frequent invitations to "add" or "substitute" various elements, emphasizing consumer choice (Jurafsky et al. 2016).

In one way, the Jordan Pond House reveals its own provenance, as the park-based outpost of a growing food-service company as opposed to the domain of, say, a buzzed-about chef: instead of being subtle and secure in its identity, it appears anxious and overhypes its food. The menu creators go completely overboard on all types of adjectives, from the participial (*caramelized* leek quinoa, *blistered* corn, *roasted* tomatoes, *candied* walnuts) to the extreme positive (*Yummy*! *Amazing*! *Refreshing*!). Liberman's 2004 work reminds us that this ostentation betrays a concern with status that would not be there in a fine-dining establishment secure in its identity.

The website for Jordan Pond House beckons with a hero image of tables on a lawn with flowers in the foreground and the Bubbles in the distance as it announces, "Tea and Popovers on the Lawn—Learn More" (Dawnland LLC 2021a). The site's narrative proclaims the restaurant's long tradition in Acadia and emphasizes the "beautiful lawn" and "amazing views" as it recommends partaking in afternoon tea with popovers.

Unlike the Metate Room in Mesa Verde (discussed later in this chapter), Jordan Pond House fails to reference local Indigenous cultures either in its food offerings or in its décor; instead, the emphasis is on the local *post* settler contact, as lobster, blueberries, and Maine draught beers are called out. This absence is intriguing, considering Dawnland LLC's laudable arts-related Tribal initiatives, including funding for Tribal partnership events and programs, Tribal art scholarships, and startup costs for Maine Tribal artists (Levin 2014).

Messaging from the outside community about health and sustainability at Jordan Pond House has been mixed. Despite its public-facing stance as deeply committed to healthy and sustainable food options, New Mexico-based Ortega came under fire for shortcomings in these areas in 2013 when it won the $6 million annual, ten-year concession contract for Acadia National Park out from under local favorites The Acadia Corp., which had held the contracts for over 80 years. Snubbed, The Acadia Corp. maintained that some of Ortega's menu items were "not obtainable or illegal in Maine or New England," including "roe from a threatened or endangered fish," thus rendering them antithetical to sustainability (Repanshek 2019). However, Ortega's subsidiary, Dawnland LLC, also received positive coverage of its promises to source all its dairy, cheese, and seafood from Maine within ten years and to radically increase the percentage of its meat and produce sourced in Maine and New England (Levin 2014).

One seemingly ironic question is whether smaller companies that lack the industrial scale of larger multi-park concessioners can deliver on the promise of sustainability. The President of the Acadia Corporation testified to a congressional subcommittee in 2015 about the challenges facing small, single-park concessioners like his own company: "Absent congressional action, national park concessions are destined to be left to companies large enough to have personnel dedicated to proposal development and centralized management offering a homogenous, mediocre service lacking the distinctiveness befitting America's unique national parks" (Woodside 2015, 2).

Users of restaurant review sites tend to act on the information they find there (Parikh et al. 2014), suggesting that the content there wields significant influence. Likewise, social media are used to communicate about food in ways that are consequential for society and individuals (Cross 2020). Thus, these spaces of food mediation bear close examination. Beyond how Jordan Pond House curates its own meanings around health and sustainability, and how those are debated in professional media, I am interested in how the experiences of everyday eaters at Jordan Pond House are represented in social media and review sites. Let me start with review sites: I looked at TripAdvisor's 2,851 reviews, with an overall rating of 4.0/5, with 4.0 for food and service, 3.5 for value, and 4.5 for atmosphere, as well as Yelp, with 680 reviews and an overall rating of 3.5/5, and Google, with 1,441 reviews, with an overall rating of 4.1/5. Across all of the review sites, common concerns emerge: Reviewers say that when reservations were accepted pre-COVID, it was hard to get anyone to answer the phone or to get a reservation at a peak time; during COVID, with no reservations, the wait is too long. Always, they say, the restaurant is crowded, with wildly inadequate parking, and service is too rushed.

Reviewers for the most part rave over the view and the landscape, but many suggest that the food comes up short. TripAdvisor's Patricia G. says, "With our meal we each ordered a popover and these are a must if you are visiting this restaurant. They are served hot, with butter and strawberry jam. They are to die for!" (June 21, 2021). But for every Patricia G., there are others lamenting the blandness, temperature, high prices, portion size, and general lack of quality. Tim B. on Yelp, for instance, went into gory detail about poor flavors and scrawny servings before concluding, "This isn't a restaurant – it's a profit maximization factory serving up junk food. Don't waste your time or money" (July 2021). Another recent Tripadvisor review, by 826bonnies, titled "Overhyped and Mediocre," sums up many of the feelings present on review sites:

> It's a head scratcher why this place is considered iconic and a must-do on an Acadia itinerary. The wait times are long and the food was quite mediocre. The fabled popovers were cold and chewy, served with paltry amounts of jam and butter. The building itself is unremarkable for National Park architecture, having a shabby 1980s summer camp in the Berkshires aesthetic. The views from the lakeshore are stunning and the reason worth stopping. Do yourself a favor and bring a boxed lunch from town to have a great picnic and avoid the restaurant letdown.
>
> *(826bonnies, Tripadvisor, June 2, 2021)*

This theme, of landscape over foodscape, exists in social media as well. Jordan Pond House Facebook posts over the last several years say little to nothing about the food; pre-COVID, it's all about the park's physical beauty, except for a rare popover closeup. Summer 2018 is the last time there were a few consecutive posts featuring the restaurant's food. Recent posts are all landscape, tables on the lawn, and Yo-Yo Ma's surprise visit ("paying tribute to the Wabanaki tribes in Maine in advance of Interior Secretary Deb Haaland's visit to the park on Friday").

In contrast to the review sites, on social media, there is little criticism of the restaurant or its food; instead, there is the replication of visual iconography. On Instagram, a recent search for the geotag "Jordan Pond House" revealed, among its top 12 posts, 5 popover photos, including the iconic Excalibur-style popover sundae with a knife sticking out. The similar geotag "Jordan Pond House" with a road address included revealed more popover shots and just as many pics of the iconic great lawn/outdoor dining setup; beyond the top posts, the others are mostly of scenery, not food. Generally, among those who use Instagram hashtags or geotags to indicate Jordan Pond House location, there is little criticism of the food or the popovers. One might get eatingwithziggy, who says "The famous Jordan Pond popovers. Not as thrilling as I remember, but quite good" alongside beauty shots of popovers and jam, but mostly, commentary is along the lines of "this is a popover!" Beyond popovers, eaters tend not to use social media to represent the rest of the food the restaurant offers, instead focusing on the scenery. Most posts that use Jordan Pond House hashtags or geotags depict the landscape and views from the restaurant, rather than the food. Many, many posts show popovers foregrounded against Jordan Pond itself. There is little commentary on the food at all, other than to clarify that the Instagrammer is participating in the ritual consumption of the iconic Jordan Pond House popover.

In my analysis of review sites and social media posts, which indicate how eaters are responding to their experiences at Jordan Pond House, there is little to no mention of sustainability, showing perhaps a misalignment of eater concerns with NPS and concessioner concerns. In this misalignment, I hear the continued drumbeat of the Romantic critique of industrialism: Eaters want hassle-free aestheticized experiences—beautiful food in beautiful settings, and although the concessioner gestures toward authenticity and distinction, reviewer critiques suggest that the industrial food service model cannot deliver on this promise. Digital media play a role in overriding critiques of the industrial food system, as the "profit maximization factory serving up junk food" in the parks, mentioned by one Yelp reviewer, becomes eclipsed by the toothless vocabulary of visual iconography that is present in social media. The social media fetishization of popover and landscape iconography harkens back to the Romantic concern for aestheticized experience, as the preoccupation with distinction manifests differently in a visually mediated reality.

Metate Room

Metate Room is the award-winning fine-dining option at Mesa Verde in southwestern Colorado. Run by Aramark, the industrial food services corporation with a 20-country footprint and concession contracts in many of the largest US national parks, Metate Room offers diners a sweeping view of the park's landscape and wildlife, including wild mustangs. Its name, Metate, evokes the concave stone against which corn was ground in Mesoamerican cultures, suggesting an awareness of and appreciation for (or appropriation of) Indigenous foodways on the part of Aramark.

Aramark opts for a minimalist approach in its web representation of Metate Room. A higher-level web page summarizes "Mesa Verde Dining Experiences," and Metate Room is one of four venues teased there with thumbnail photos and enticing taglines. Headlined "A Feast for All Your Senses," the page offers the possibility of an "elegant evening" with "breathtaking views" and notes that "the menus at Mesa Verde National Park celebrate the local culture with favorites inspired by the past and present" (Aramark 2022). A thumbnail photo depicts a completely unoccupied Metate Room, with tables and chairs in the foreground and the floor-to-ceiling wall of windows looking out onto the scrub of Chapin Mesa in the background. Beneath the thumbnail, a teaser reads, "Colorful Native American artwork surrounds you as you enjoy our award winning, contemporary menu inspired by regional heritage foods and flavorings."

Upon clicking through to Metate Room's own web page, one finds a slider with eight photos depicting the empty dining room, Native American artwork on the walls, and lovely views, a table set for two with wine chilling and artfully arranged food on plates, and aesthetically pleasing close-ups of ornately plated salmon, steak, salad, and a brownie.

The pictures of food are artful, well-lit, and competently composed shots with an occasional human arm or hand visible. Beyond the photos, the text is sparing. Readers are advised of the hours of operation and how to make a reservation, which is strongly recommended, and the line about colorful Native American artwork is repeated, but no other claims are made about the restaurant. Links are provided to three menus: dinner, kids, and dessert.

Like the website, the dinner menu for Metate Room does little to make claims about health or sustainability. It does feature icons to indicate vegetarian or gluten-free options, and a box at the bottom of the menu provides a stock warning about consuming raw or undercooked foods; it also advises that additional nutrition information is available upon request and suggests that 2,000 calories a day is a good general guideline. This connects to the calorie information that accompanies each menu item listed. There are no claims to organic, healthy, or local food on the menu.

Metate Room's menu, with starters ranging from $6.75 to $15, salads from $6 to $12, and entrees from $24 to $35, seems to indicate reasonable prices. As we saw at Jordan Pond House, many of the modifiers describing the preparation of each item emphasize authenticity, tradition, abundance, or choice: The focaccia crisp is "house-made" and the chicken wings are "house seasoned," the charcuterie board comes with "traditional accompaniments," the arugula and the cheese are "fresh," the au jus is "natural," and the angus beef is "hand-carved." But the overall tone of the Metate Room is more subdued, with less of the anxious, overhyped tone seen on some other park restaurant menus. The menu offerings are more adventurous than is typical at park restaurants, with nary a burger in sight. Although familiar options of beef, chicken, and fish are represented, there are some preparations that stray from the expected: The chicken wings are "Korean BBQ" style, prepared with gochujang BBQ sauce alongside the typical celery and blue cheese. There are multiple dishes that feature ancient grains. And there are a handful of offerings containing elements—crème fraiche, candied pancetta, herbed red wine demi-glace, red pepper aioli, goat cheese, and caper sauce—that are upscale, allowing cosmopolitan foodie visitors to feel at home. How the promised "regional heritage foods and flavors" find expression on the menu is anyone's guess, as there is no indication of local or regional sourcing, nor are brand names of any ingredients provided.

Messaging from the outside professional community about Metate Room tends to praise the quality of the food and to highlight the creativity and talent of its executive chef. An article on the National Parks Traveler site from 2010 applauds Metate Executive Chef Brian Puett's win of the "Award of Culinary Excellence" by the American Culinary Federation Colorado Chefs Association. The award acknowledged Metate Room's involvement with the Colorado Department of Agriculture's program to promote Colorado food (Repanshek 2010). The National Parks Traveler article notes Puett's efforts at "creating contemporary interpretations of heritage foods from the original Mesa Verde inhabitants, while maintaining ARAMARK's commitment to incorporating sustainable and organic ingredients in all menus" (Repanshek 2010), which is fascinating praise, given the lack of information about sustainable, organic, local, or heritage elements in the way Aramark itself presents Metate Room. An earlier feature from 2009 in the same publication profiles Chef Puett and emphasizes his reliance on heritage foods of the Southwest—bison, elk, turkey, quail, squash, black beans, tortillas, and prickly pears—in building his menu (Repanshek 2009). A 2015 feature in the *Durango Herald* commends Puett's successor, Derek Fontenot, for his formal training in classical French cuisine; it notes that "He has been working to increase the number of fresh, local and sustainable offerings, not an easy feat when food delivery trucks make the trek up the switchbacks only twice a week" (Anesi 2015). The author acknowledges the Metate Room's Southwestern regional influence, citing the presence of squash, beans, and corn alongside Hatch green chilis, poblanos, prickly pear, pine nuts, and red pepper coulis (Anesi 2015). Similar themes are echoed in a piece on MesaVerdeCountry.com: Fontenot's French cooking expertise is foregrounded, supported by details about his commitment to "getting products locally whenever possible, to support the regional economy and a sustainable farm-to-table ethos" (Mesa Verde Country, n.d.).

At the time of this writing, in 2022, the Metate Room is seeking an executive chef, and most traces of these regional specialties or evidence of local sourcing are absent from the menu. In 2009, Chef Puett was quoted as saying that Aramark requires him to get 80% of his food ingredients through its national distributor, but that he tries to buy locally for greater sustainability and to support smaller farmers (Repanshek 2009). One can only speculate that such conditions are constricting for talented, creative chefs, while noticing that today's menu is more pro forma than the versions profiled a decade ago. The menu stands in striking contrast to the focus on local and Indigenous foods at the Mesa Verde Museum Association Bookstore in the Mesa Verde visitor center just down the road, where Indigenous foods (bags of blue cornmeal, tiny burlap sacks of Anasazi beans, and Cliff Dweller bean soup mix) and books about Indigenous foodways (Frank's *Foods of the Southwest Indian Nations*, Dahl's *Native Harvest*, Niethammer's *American Indian Cooking*, and Swentzell and Perea's *The Pueblo Food Experience Cookbook*) are the first things a visitor sees upon entering the store.

Professional community commentary on Metate Room takes up discourses of sustainability and status, but not health. But what do Metate Room eaters have to say about their experiences there? To discern what discourses prevail in the eater imaginary, I examined Tripadvisor's 712 reviews, with an overall rating of 4.0/5, with 4.0 for food and service, 3.5 for value, and 4.5 for atmosphere, as well as Yelp, with 80 reviews and an overall rating of 3.5/5, and Google, with 51 reviews, with an overall rating of 3.8/5. Across all the review sites, it is difficult to find consistency, except for commentary on the views (sometimes including horses on the horizon) from the Metate Room. Opinions on the food position it as ranging from out-of-this-world delicious to terrible, but overall, reviewers seem to like the flavors and presentation. In general, reviewers argue that the food is anywhere from a little expensive to ridiculously high-priced. At the same time, they acknowledge that there are few alternatives nearby; the words "captive" and "monopoly" show up often. Many reviewers begrudgingly give props to Aramark, stating that they had low expectations of an industrial concession going in, but that the dining experience was surprisingly good. Reviews of service are quite varied; there are frequent complaints about slow or poor service, but just as many comments about the friendliness and skill of the servers. Several reviewers claim to have eaten in the high-end restaurants of many national parks, thus establishing their credibility as judges, before stating that Metate Room is the best or one of the best.

The theme of landscape over foodscape that dominates internet reviews of Jordan Pond House is less pronounced in reviews of Metate Room, where reviewers seem to have a more positive perspective on the culinary offerings. Some reviews, like the one posted by Jeffrey Cole on Google Reviews in September 2021, evaluate both positively: "The restaurant has a stunning wraparound view of the mesa and the distant San Juan mountains, so beautiful. The menu is limited but 'curated,' more suited to adults with expensive tastes." An effusive Yelp review, posted by Jon L. in August 2018, mentions the sunset views only to explain why the service was off during an otherwise spectacular dining experience:

> Better than Tom Colicchio, blew away Thomas Keller. We were driving back from Vegas where we were doing the foodie tour of our favorite chefs' restaurants. This review isn't about those places, but the Metate Room was the best meal we had on the entire trip. As brilliantly executed as the classic steakhouse experience was in Vegas the Metate Room brought real creativity and a modern twist to the entrees. A perfect approach to traditional ingredients with just the right amount of southwest flair. The service struggled a bit, but I understand why. It's the Far View Lodge and everyone wants to camp out until the fantastic sunset. It's not their fault that they were overloaded, but they always stayed polite and the food was fantastic.
>
> *(Jon L., Yelp, August 26, 2018)*

Others are less enthusiastic about the food without ever mentioning the view, like Chuck: "It may be masquerading as a fine restaurant, but it's still run by a giant food service corporation, and it shows. Bad food, high prices" (Chuck L., Yelp, July 9, 2017). Other Yelp critics describe the menu descriptions as "typically dramatic" but claim that "the actual product is not even close," suggesting that the restaurant does not even meet the low bar of tolerability set by its remote location (John E., Yelp, May 18, 2021).

Overall, reviews posted on these sites focus more on the food than on the landscape, and many comments about food assert a cosmopolitan identity on the part of the reviewer. Steph D. asserts that her meal at Metate Room is "the MOST impressive National Park food I have ever had," describing it as authentic and well-prepared. She provides details on price, ambiance, service, and views, but most of her review describes the food, noting that the local "native Mesa Verde ingredients" of corn, beans, squash, and quail were very impressively used in what her party ate. She summarizes by assuring readers that what they'll get at Metate Room is of equal quality to urban high-end dining: "The quality of the food is in line with major city fine dining…think NYC, Miami" (Steph D., Yelp, August 5, 2012). Other reviewers reveal their evaluative capacities by describing the flavor of the dish in depth, like VeryVito on Tripadvisor who waxes poetic about his pork, "the best meal I've ever had at a restaurant" (July 13, 2021), and mikeg-C6777EY, also on Tripadvisor, who describes the food as "surprising [sic] good for a corporate venue" (July 12, 2021). This theme is picked up by SusanH2466 on Tripadvisor, who admits, "Our expectations for concessionaire dining is [sic] pretty low, however the Metate Room provided a fabulous meal." She acknowledges that the small number of items on the menu allowed the chef to concentrate on a few items, an endeavor she deems successful before concluding that "we left with much higher expectations of other restaurants in the national park system" (September 20, 2020). Even when reviewers describe the food as just "meh," as millukeer418 did on Tripadvisor, some position themselves as discerning foodies not easily taken in: "The Metate Room is touted as the fine dining option for foodies who visit the park," he writes, while noticing that the much-vaunted reservations seemed unnecessary given that the dining room was only a third full during his visit; he concludes, "Maybe to drum up excitement for an otherwise mediocre restaurant?" (millukeer418, Tripadvisor, August 29, 2019). These reviewers demonstrate cosmopolitan traits—well-traveledness and an ability to render assessments of quality, whether euphoric, passable, or poor.

Review site postings are limited for Metate Room, compared to Jordan Pond House. Likewise, the social media presence is quite limited as well. On Instagram, there are only 25 posts hashtagged #metateroom and just a half dozen that use the Metate Room geotag. Among the posts that do exist, there is an equal balance of view and food shots, and there are a handful of images that combine the two, with the table/food in foreground and the view in background. Some of the photos are of the Metate Room's décor, and a few are of the user's dining companions. Just as there is no menu equivalent here of the famous Jordan Pond House popover, there is no iconic food shot, perhaps because Metate Room is far from the actual cliff dwellings that are visually distinctive. On Instagram, posts with Metate Room hashtags or geotags do not provide commentary on health or sustainability, but here there are some beautiful pictures of food.

Jenny Lake Lodge Dining Room

Grand Teton Lodge Company (GTLC), owned by Vail Resorts Hospitality, is one of two food service concessioners in Wyoming's Grand Teton National Park. (The other food concessioner, Signal Mountain Lodge, is owned by Forever Resorts, which is a subsidiary of Aramark.) Vail Resorts Hospitality is a large corporation that owns or manages destination resorts, golf courses, and ground transportation. Within Grand Teton National Park, the Jenny Lake Lodge Dining Room (JLLDR) is the jewel of GTLC's portfolio—the upscale restaurant at its exclusive and "rustically elegant retreat" with just 37 luxury cabins (Vail Resorts Hospitality 2022). Guests at the "only 4-diamond eco-resort in the Park" who "seek the finest in service and lodging" receive, included in their very expensive (close to $1,000 per night for two people) stay, a daily "gourmet" breakfast and five-course dinner at the JLLDR (Vail Resorts Hospitality 2022). Would-be diners who are not staying at the Jenny Lake Lodge can take their chances on lunch or dinner reservations on OpenTable, but availability is limited.

Despite its jewel status, the JLLDR is not particularly promoted above other GTLC-run restaurants on the GTLC Dining web page. However, GTLC makes its commitments to sustainability and quality known from the outside, with text front and center:

> Dining in Grand Teton National Park is not what many have some to expect of National Park fare. Each of our specialized dining outlets offers tantalizing local dishes that are as unique as the experience of visiting the park itself. Our Executive Chef Joshua Gayer and his professional culinary staff carefully prepare each dish using locally grown organic produce, free range and hormone free meats, and sustainable seafood abiding by the guidelines of the Monterey Bay Aquarium Seafood Watch program. Your dining experiences at our properties will reflect hearty regional cuisine at its finest.
>
> *(Grand Teton Lodge Company 2022a)*

Just above, large slider photos in the hero position on the site depict three other dining venues in the park. Scrolling down, one finds more than a dozen clickable tiles with thumbnail photos of each GTLC eatery. JLLDR's tile teases "The Crown Jewel of Fine Dining in Grand Teton National Park" and shows a table for two set with white tablecloth, a cute vase of yellow cinquefoils and pine needles, and stemmed glasses of white wine, up against a window looking out to a meadow, trees, and the breathtaking mountains. One can find out more about this restaurant labeled "$$$$" by clicking through to the JLLDR page, where one is greeted by a larger version of the thumbnail photo and prominent icons indicating the restaurant's many awards— among them, Tripadvisor Traveler's Choice 2020, Tripadvisor Certificate of Excellence 2015–2019, and Wine Spectator Award of Excellence 2020.

GTLC makes grand claims on the JLLDR site, reflecting a focus on status and sustainability:

> Housed inside a 1930's-era log cabin nestled in the woods at the base of the Teton Range, this AAA Four-Diamond dining destination offers breakfast, lunch and a nightly rotating five-course prix fixe dinner menu that justifies its place among the world's finest restaurants. Good for families and couples alike,

The Dining Room offers a culinary experience unlike any other found inside a national park. It is so special that many make dining here an annual event. The Jenny Lake Lodge culinary team locally sources high-quality sustainable ingredients, with many coming from the surrounding Jackson Hole area. Their mouthwatering creations, combined with an award-winning wine list, results in a dining experience that delights even the most refined palates.

(Grand Teton Lodge Company 2022b)

The web page effuses distinctiveness: The dining experience at JLLDR is like no other in a national park; it's primarily for the refined insiders who go back year after year and can afford a room there, and it's about quality and sustainability.

The web page also contains links to four sample dinner menus from the 2021 season; in the tradition of old-school high-end restaurants, no prices are listed, although this is understandable given that the dinner experience is clearly framed on the website as a five-course tasting menu costing $115 per person. The four menus signal to knowledgeable, cosmopolitan diners, with plenty of "continentalized" non-English words that are likely to be unfamiliar to those who don't actively pursue food-related education (Jurafsky 2014; Krystal 2018). Feeling comfortable with JLLDR menu terms like compote, gelee, rocket, carpaccio, crudo, ragout, duxelles, velouté, poussin, fonduta, coulis, and affogato is likely to leave in-the-know eaters feeling satisfied with their culinary erudition. The very decision to include these terms on the menus may be part of a restaurant industry strategy to increase descriptive complexity to communicate upper-tier quality (McCall and Lynn 2008). Another element of descriptive complexity on the menu is the presence of relatively few adjectives connoting action (whipped, housemade, pickled), outnumbered by more nouns simply naming the ingredients. Culinary trends from Michelin-starred restaurants, like a dish of "raviolo" (note the singular), find their way onto the menu (Grand Teton Lodge Company 2021b). In a few cases, the menus present a kind of chef-fy shorthand, like in a dish of "White Tail Venison Medallion" with "cherry demi, duroc-heirloom potato hash" (Grand Teton Lodge Company 2021b). If you know, you know, and you do not need "demi-glace" spelled all the way out. Likewise, you do not need to look up "duroc" to recognize it as a popular US pig breed. The menus here employ culinary enthymemes that rely on the knowledge of the educated cosmopolitan consumer to do interpretive work.

Another way that the menus at JLLDR hail cosmopolitan diners is by their unusual-for-a-national-park insistence on naming the executive chef. As Lane (2013, 2014) has written, buzz about a vaunted chef provides symbolic value that enhances diners' perception of their gustatory pleasure (see also Rousseau 2012), and this same emphasis on the chef is absent in casual dining experiences where the cooks are understood to be interchangeable. On the 2021 JLLDR menus, the Executive Chef is identified mononymically, like Cher or Madonna, as *"Nonelius,"* suggesting a certain level of mythmaking at work.

2021 JLLDR menus deal with local sourcing in a somewhat obfuscatory manner. At the bottom of each menu, just after a statement about the potential dangers of consuming raw or undercooked foods, and advice to let servers know about dietary restrictions, is a noncommittal statement: "All menu items prepared using local and sustainable products as available" (Grand Teton Lodge Company 2021a). All bets are hedged as the GTLC provides little further detail on the menu or website itself about the venues from which it buys its food. There are several dishes on the 2021 menus that nod to the local, like the "Jenny Lake Signature Salad" that includes house-made farmer cheese, Roasted Whole Idaho Trout, and a cheese board for dessert that includes "Chef selection of locally sourced cheese" (Grand Teton Lodge Company 2021a). Beyond these somewhat vague references, several menus boast place-based items, like the Duck Poached Skuna Bay Salmon and the Hudson Valley Foie Gras Mousse (Grand Teton Lodge Company 2021b), and the Colorado Striped Bass (Grand Teton Lodge Company 2021c). While all these labels signify provenance, none of them are exactly local to western Wyoming. In this case, diners are provided appealing, place-based descriptive information that nonetheless makes no commitment to sustainable or local food sourcing.

Compared to Metate Room and Jordan Pond House, there is a paucity of coverage of the JLLDR in professional media. Despite its strong reputation as a distinctive dining venue in the national park system, there is less of a presence of food blog or even newspaper coverage of JLLDR. It is mentioned in a review blurb fashion on travel sites like Fodor's Travel, Lonely Planet, and Frommer's, which collectively emphasize its surprisingly high-quality food, local sourcing, elegantly rustic setting, and ambitious and inventive menu while warning would-be diners that reservations are necessary. Beyond that, there is little to be found, which may signify a less aggressive public relations strategy on the part of GTLC than anything else.

Luckily, even in the absence of much professional community commentary on JLLDR, one can glean how the key themes communicated by the restaurant find expression in the minds of consumers by looking at review sites. I examined Tripadvisor's 401 reviews of the restaurant, with an overall rating of 4.5/5, with 4.5 for food, service, and atmosphere, and 3.5 for value, as well as Yelp, with 45 reviews and an overall rating of 4.0/5, and Google, with 17 reviews, with an overall rating of 4.1/5, to see what diners had to say about their experiences there. On the whole, users reviewing JLLDR say nothing about health, and very little about sustainability, except to occasionally repeat JLLDR's claim of local sourcing (something that appears to have been communicated about more clearly on menus prior to the 2021 season, to judge by the photos some reviewers have attached). Emphasis is squarely on setting, taste, and service.

Review site users rate Jenny Lake Lodge highest for food of any national park restaurant that I have examined and take care to point out the scenery, while lovely, is not the main attraction. On Tripadvisor, for instance, aut0lysej writes, "Best food in a national park. Most of the restaurants in the national parks are run by concessionaires with scant attention to the food. The atmosphere is supposed to be worth the meal. Not so here. This is the best food I have ever had in a national park--inventive and tasty. And the service was excellent, not to mention that the log building itself is a perfect place to dine" (aut0lysej, Tripadvisor, September 1, 2019). This point is echoed on Yelp, where Margaret B. writes, "The food is really what shines here; just happens to be coupled with exquisite views of the Teton mountains and wonderful service" (Margaret B., Yelp, August 2, 2017).

More than anything, the review sites for JLLDR function as spaces for the negotiation of status. Although the reviews are overwhelmingly positive, there are a handful that rail against the restaurant for prioritizing overnight guests of the (very expensive) lodge, or that bemoan the unaffordable prices. On Tripadvisor, Senior-TravelerTF warned, "If you aren't a Food Network junkie don't spend the money," suggesting that he'd only recommend the restaurant to those who "can name the top chefs of New York instead of the starting lineup for the Yankees" (SeniorTravelerTF, Tripadvisor, September 28, 2018). This user identifies as *not* a foodie, but his contention that this restaurant is a foodie destination is shared by others who take their status as foodies more seriously. RevLex101 titles their post "Gourmet Food with a Stunning View," and writes, "I've eaten at high-end restaurants all over the world and would place this little gem near the top of the list. The menu changes daily, and the food is amazing" (RevLex101, Tripadvisor, September 27, 2018).

Review site reviews of JLLDR are somewhat unique in that there is a fair amount of explicit discourse about foodie-ism. From the first review of JLLDR posted on Tripadvisor, by Robert K., one learns that it is "a must for any foodies visiting the Tetons" (Robert K., Tripadvisor, July 8, 2011), but later, David L. praises the setting and view but suggests that foodies will be dissatisfied with the actual flavors:

> The food was good, but has slipped since our previous visit in 2016, when it was outstanding. The service was attentive, but a bit overly dramatic and chatty, as opposed to typical fine dining establishments. The bottom line is that Jenny Lake Lodge is still an amazing place, but not a foodie destination this year.
>
> *(David L., Tripadvisor, July 28, 2018)*

The very cosmopolitan capacity for critique and discernment is well evidenced on the review sites and finds itself expressed in the theme that the JLLDR is out over its skis. Charliespal2 critiques the restaurant as "trying too hard to be gourmet" (Charliespal2, Tripadvisor, August 29, 2018), while Maria S. offers more color commentary on the off-putting nature of the restaurant's pretension:

> The food server was friendly, but the whole thing was rather pretentious and since the food was mediocre, it was more annoying than helpful to be given a play by play of the ingredients and preparation of each course we were served. The whole thing was a big disappointment and a costly one at that.
>
> *(Maria S, Tripadvisor, September 8, 2018)*

Other posters, without ever using the word "foodie," flex their evaluative capabilities in discussing the food. Describing his dessert choice, FlyWagon writes, "I indulged in an exquisite glass of Far Niente's 2012 "Dolce", a Late Harvest Sauvignon Blanc, the closest I've had to a true Sauternes in years" (Tripadvisor, September 27, 2018). There is no doubt that this poster knows his wines well enough to be able to assess what a "true" Sauternes would be and to evaluate the Dolce as "exquisite," a more highfalutin' word than "good." Similarly, Sara C. indulges in a detailed analysis of the inconsistencies of the dishes she and her family tried:

> My husband's pork cheek entree was tender and delicious. But it was served with a relatively bland foie gras bread pudding and half of a plum oddly on the side. Several dishes came with cilantro. Maybe this is just my pet peeve, but whenever a dish has a strong polarizing ingredient like that, I think it should be listed, especially if it's in a dish you wouldn't typically expect cilantro to appear in. (Meanwhile, they would list benign ingredients like parsley). Also found it a bit strange that they featured so many non local ingredients. Are they really known for octopus? The standout dish to me was the artichoke heart entree. Truly delicious. And the homemade focaccia bread was pretty good.
>
> *(Sara C., Tripadvisor, August 11, 2017)*

There's nothing exactly haughty in Sara C.'s post, but her judgments about the plum ("odd"), the cilantro ("strong polarizing"), parsley ("benign"), and the preponderance of nonlocal ingredients (octopus) demonstrate that she is educated and well versed in culinary discourses that bring status to those who participate in them.

JLLDR seems to have largely escaped notice on Instagram, where it has been hashtagged in only 11 posts and geotagged in just 2. In addition, almost all the hashtagged posts are teasers for a 2019 film, *Send It*, about the restaurant and its local sourcing efforts. The geotag for the JJLDR reveals only two posts—one showing a group of older patrons and the other showing a plate and napkin setting. The GTLC's contention that JJLDR provides a unique and distinctive dining experience is not borne out in the restaurant's social media presence, where neither the company nor those who dine there have contributed much to its visual representation.

Conclusion

High-end park eateries like the JLLDR, Metate Room, and the Jordan Pond House employ various representational strategies to reach the public. Some emphasize discourses of health and sustainability, while others highlight distinctiveness and quality. In each case, professional media and consumers both reinforce and resist corporate messaging in interesting ways that show what cosmopolitan eaters expect from eating experiences in national parks. Although there is a temptation to dismiss these cosmopolitan desires, as Chapter 6 demonstrates, their interest in environmental sustainability, artistanality, and social connections across difference may be an important catalyst for future park food ecologies.

References

Anesi, Karen Brucoli. 2015. "Dining Among the Dwellings: Metate Room is Worth the Trip Up the Mesa." *The Durango Herald*, September 10, 2015. http://api.the-journal.com/articles/27330.

Appadurai, Arjun. 1990. "Disjuncture and Difference in the Global Cultural Economy." *Theory, Culture, and Society* 7 (2–3): 295–310.

Aramark. 2022. "Mesa Verde Dining Experience." Accessed August 21, 2022. https://www.visitmesaverde.com/dining/mesa-verde-dining-experience/.

Bell, David and Joanne Hollows. 2007. "Mobile Homes." *Space and Culture* 10 (1): 22–39. https://doi.org/10.1177/1206331206296380.

Clayton, John. 2019. "Modernizing National Park Facilities: Mission 66 in Wyoming." *WyoHistory.org*, September 18, 2019. https://www.wyohistory.org/encyclopedia/modernizing-national-park-facilities-mission-66-wyoming.

Contois, Emily J.H. and Zenia Kish, ed. 2022. *Food Instagram: Identity, Influence, and Negotiation*. Urbana-Champaign: University of Illinois Press.

Cross, Karen. 2020. "Visioning Food and Community through the Lens of Social Media." In *Digital Food Cultures*, edited by Deborah Lupton and Zeena Feldman, 162–76. London: Routledge.

Dawnland LLC. 2021a. "Jordan Pond House." Accessed July 21, 2021. https://jordanpondhouse.com/jordan-pond-house/.

Dawnland LLC. 2021b. "Jordan Pond House Menu (Downloadable)." Accessed July 20, 2021. https://jordanpondhouse.com/wp-content/uploads/2021/07/JPH-2021-Menu.pdf.

Feldman, Zeena. 2021. "'Good Food' in an Instagram Age: Rethinking Hierarchies of Culture, Criticism and Taste." *European Journal of Cultural Studies* 24 (6): 1340–59. https://doi.org/10.1177/13675494211055733.

Feldman, Zeena and Michael K. Goodman. 2021. "Digital Food Culture, Power and Everyday Life." *European Journal of Cultural Studies* 24 (6): 1227–42. https://doi.org/10.1177/13675494211055501.

Figueiredo, Bernardo, Hanne Pico Larsen, and Jonathan Bean. 2021. "The Cosmopolitan Servicescape." *Journal of Retailing* 97 (2): 267–87. https://doi.org/10.1016/j.jretai.2020.09.001.

Grand Teton Lodge Company. 2021a. "Jenny Lake Lodge Menu 1." Accessed August 29, 2022. https://www.gtlc.com/media/3069/jny-menu-1-2021.pdf.

Grand Teton Lodge Company. 2021b. "Jenny Lake Lodge Menu 2." Accessed August 29, 2022. https://www.gtlc.com/media/3072/jny-menu-2-2021.pdf.

Grand Teton Lodge Company. 2021c. "Jenny Lake Lodge Menu 3." Accessed August 29, 2022. https://www.gtlc.com/media/3071/jny-menu-3-2021.pdf.

Grand Teton Lodge Company. 2022a. "Dining." Accessed August 26, 2022. https://www.gtlc.com/dining.

Grand Teton Lodge Company. 2022b. "Jenny Lake Lodge Dining Room." Accessed August 26, 2022. https://www.gtlc.com/dining/the-dining-room-at-jenny-lake-lodge.

Grusin, Richard. 2004. *Culture, Technology, and the Creation of America's National Parks*. Cambridge: Cambridge University Press.

Gualtieri, Gillian. 2021. "Discriminating Palates: Evaluation and Ethnoracial Inequality in American Fine Dining." *Social Problems* 69 (4): 903–927. https://doi.org/10.1093/socpro/spaa075.

Hannerz, Ulf. 1996. *Transnational Connections: Culture, People, Places.* London: Routledge.

Hansen, Christine. 2015. "The Future Fault Lines of Food." In *The Future of Food Tourism: Foodies, Experiences, Exclusivity, Visions and Political Capital,* edited by Ian Yeoman, Una McMahon-Beattie, Kevin Fields, Julia N. Albrecht, and Kevin Meethan, 49–61. Bristol: Channel View.

Herman, Andrew. 2021. "Dynamic Status Signaling: How Foodies Signal Cosmopolitanism on Yelp." *Poetics* 90 (3): 101592. https://doi.org/10.1016/j.poetic.2021.101592.

Johnston, Josée and Shyon Baumann. 2007. "Democracy Versus Distinction: A study of Omnivorousness in Food Writing." *American Journal of Sociology* 113 (1): 165–204.

Johnston, Josée and Shyon Baumann. 2015. *Foodies: Democracy and Distinction in the Gourmet Foodscape.* Second Edition. New York: Routledge.

Jurafsky, Dan. 2014. *The Language of Food: A Linguist Reads the Menu.* New York: W.W. Norton and Co.

Jurafsky, Dan, Victor Chahuneau, Bryan R. Routledge, and Noah A. Smith. 2014. "Narrative Framing of Consumer Sentiment in Online Restaurant Reviews." *First Monday* 19 (4). https://doi.org/10.5210/fm.v19i4.4944.

Jurafsky, Dan, Victor Chahuneau, Bryan R. Routledge, and Noah A. Smith. 2016. "Linguistic Markers of Status in Food Culture: Bourdieu's Distinction in a Menu Corpus." *Journal of Cultural Analytics* 1 (1): 1–24. https://doi.org/10.22148/16.007.

Kobez, Morag. 2020. "A Seat at the Table: Amateur Restaurant Review Bloggers and the Gastronomic Field." In *Digital Food Cultures,* edited by Deborah Lupton and Zeena Feldman, 99–113. New York: Routledge.

Krystal, Becky. 2018. "Foreign Words Often Used to Fancy Up Menus at Restaurants." *Arkansas Democrat Gazette.* Posted February 11, 2018. https://www.arkansasonline.com/news/2018/feb/11/foreign-words-often-used-to-fancy-up-me/.

Lane, Christel. 2013. "Taste Makers in the 'Fine-Dining' Restaurant Industry: The Attribution of Aesthetic and Economic Value by Gastronomic Guides." *Poetics* 41 (4): 342–65.

Lane, Christel. 2014. *The Cultivation of Taste: Chefs and the Organization of Fine Dining.* New York: Oxford University Press.

Leer, Jonatan. 2019. "Monocultural and Multicultural Gastronationalism." *European Journal of Cultural Studies* 22 (5–6): 817–34.

Levin, Robert. 2014. "Contract Brings Concession Changes to Acadia." *The Ellsworth American,* July 6, 2014. https://www.ellsworthamerican.com/news/contract-brings-concession-changes-to-acadia/article_682240e8-576f-5381-9254-4ff9e6417120.html

Lewis, Tania. 2020. *Digital Food: From Paddock to Platform.* New York: Bloomsbury Academic.

Liberman, Mark. 2004. "Modification as Social Anxiety." *Language Log,* May 16, 2004. http://itre.cis.upenn.edu/~myl/languagelog/archives/000912.html.

Litvin, Stephen W., Julia E. Blose, and Stephen T. Laird. 2005. "Tourists' Use of Restaurant Webpages: Is the Internet a Critical Marketing Tool?" *Journal of Vacation Marketing* 11 (2): 155–61. https://doi.org/10.1177/1356766705052572.

Lupton, Deborah. 2016. "Cooking, Eating, Uploading: Digital Food Cultures." In *The Bloomsbury Handbook of Food and Popular Culture,* edited by Kathleen LeBesco and Peter Naccarato, 66–79. New York: Bloomsbury.

Mak, Athena H.N., Margaret Lumbers, and Anita Eves. 2012. "Globalisation and Food Consumption in Tourism." *Annals of Tourism Research* 39 (1): 171–96. https://doi.org/10.1016/j.annals.2011.05.010.

Mapes, Gwynne. 2018. "(De)constructing Distinction: Class Inequality and Elite Authenticity in Mediatized Food Discourse." *Journal of Sociolinguistics* 22 (3): 265–87. https://onlinelibrary.wiley.com/doi/abs/10.1111/josl.12285.

McCall, Michael and Ann Lynn. 2008. "The Effects of Restaurant Menu Item Descriptions on Perceptions of Quality, Price, and Purchase Intention." *Journal of Foodservice Business Research* 11 (4): 439–45. http://doi.org/10.1080/15378020802519850.

Mesa Verde Country. n.d. "Fresh Four Corners Cuisine Inspired by French Classics." *Mesaverdecountry.com*. Accessed August 25, 2022. https://mesaverdecountry.com/fresh-four-corners-cuisine-inspired-by-french-classics/.

National Park Service. 2017. "National Park Service Healthy Food Choice Standards and Sustainable Food Choice Guidelines for Front Country Operations." Last revised June 15, 2017. https://www.nps.gov/subjects/concessions/upload/Frontcountry_Healthy_Foods-2.pdf.

National Park Service. 2021. "Eating and Picnicking." Acadia National Park. Accessed July 21, 2021. https://www.nps.gov/acad/planyourvisit/eating.htm.

Parikh, Anish, Carl Behnke, Mihaela Vorvoreanu, Barbara Almanza, and Doug Nelson. 2014. "Motives for Reading and Articulating User-Generated Restaurant Reviews on Yelp.com." *Journal of Hospitality and Tourism Technology* 5 (2): 160–76. https://doi.org/10.1108/JHTT-04-2013-0011.

Repanshek, Kurt. 2009. "Dining at The Parks: Mesa Verde National Park's Chef Ensures The Southwest Flows through His Dishes." *National Parks Traveler.org*, July 18, 2009. https://www.nationalparkstraveler.org/2009/07/dining-parks-mesa-verde-national-parks-chef-ensures-southwest-flows-through-his-dishes.

Repanshek, Kurt. 2010. "Metate Room Restaurant at Mesa Verde National Park Recognized for Its Excellence." *National Parks Traveler.org*, January 24, 2010. https://www.nationalparkstraveler.org/2010/01/metate-room-restaurant-mesa-verde-national-park-recognized-its-excellence5254.

Repanshek, Kurt. 2019. "Private Equity Firm Takes Majority Interest in Ortega National Parks." *National Parks Traveler.org*, November 11, 2019. https://www.nationalparkstraveler.org/2019/11/private-equity-firm-takes-majority-interest-ortega-national-parks.

Rousseau, Signe. 2012. *Food Media: Celebrity Chefs and the Politics of Everyday Interference*. London: Berg.

Sheetz, Dakota. 2020. "A Look into Restaurant Profit Margins." *Restaurant Owner and Manager*, October 9, 2020. https://rmagazine.com/articles/a-look-into-restaurant-profit-margins.html.

Sims, Rebecca. 2009. "Food, Place and Authenticity: Local Food and the Sustainable Tourism Experience." *Journal of Sustainable Tourism* 17 (3): 321–36. https://doi.org/10.1080/09669580802359293.

Sulem, Matt. 2016. "The Best Restaurants Inside 20 Different National Parks." *TheDailyMeal.com*, April 14, 2016. https://www.thedailymeal.com/travel/best-restaurants-inside-20-different-national-parks-slideshow/slide-15.

Turnwald, Bradley P., Dan Jurafsky, Alana Conner, and Alia J. Crum. 2017. "Reading between the Menu Lines: Are Restaurants' Descriptions of "Healthy" Foods Unappealing?" *Health Psychology* 36 (11): 1034–37. https://doi.org/10.1037/hea0000501.

Vail Resorts Hospitality. 2022. "Grand Teton Lodge Company." Accessed August 26, 2022. http://www.vailresorts.com/Corp/info/grand-teton-lodge-company.aspx.

Wise, Amanda and Selvaraj Velayutham. 2009. *Everyday Multiculturalism*. Basingstoke: Palgrave Macmillan.

Woodside, David B. 2015. "Statement of David B. Woodside, President of the Acadia Corporation before the Subcommittee on Interior, Committee on Oversight and Government Reform, U.S. House of Representatives on Modernizing the National Park Service Concessions Program." July 23, 2015. chrome-extension://efaidnbmnnnibpcajpcgl-clefindmkaj/https://static1.squarespace.com/static/62d84f719b9f79362f3bb0e7/t/62f65 93b586ec975042070aa/1660311867900/Dave-Woodside-Testimony.pdf

Yilmaz, Gökhan and Selami Gültekïn. 2016. "Consumers and Tourists' Restaurant Selections." In *Global Issues and Trends in Tourism*, edited by Cevdet Avcikurt, Mihaela S. Dinu, Necdet Hacioglu, Recep Efe, Abdullah Soykan, and Nuray Tetik, 217–30. Sofia: St. Kliment Ohridski University Press.

Zeamer, Victoria Jean. 2018. "Internet Killed the Michelin Star: The Motives of Narrative and Style in Food Text Creation on Social Media." Unpublished MA thesis, Massachusetts Institute of Technology. https://dspace.mit.edu/handle/1721.1/117899.

6

REIMAGINING FOOD IN NATIONAL PARKS

Future Ecologies of Bioregionalism and Indigenous Food Sovereignty

The mandates of the 1916 Organic Act, for both conservation *and* enjoyment by the public, have generated a concession system that many eaters feel does not have its priorities straight. In the previous chapters, I have raised many concerns about the historical and contemporary state of foodways in US national parks, but I'm heartened by the possibility of forward-looking foodways. My priorities for these forward-looking park foodways are based in concern for environmental sustainability, and for those original locals, Indigenous peoples, whose own foodways were radically disrupted to create the national parks in the first place. This conclusion explores the changes in public sentiment and policy that offer a path to realizing these priorities and examines the extent to which these priorities are compatible with the project of cosmopolitan self-making.

The Appeal of Bioregionalism

Bioregionalism is the idea of becoming native to place (Andruss et al. 1990), valuing local ecological knowledge and respecting Indigenous communities. Slow food tourism is the mode of travel and exploration most closely associated with a bioregionalist framework: It "valorizes artisanal, handmade and quality local foods that offer rich aesthetic experiences and are a conduit to a sense of place and *terroir*" (Fusté-Forné and Jamal 2020, 227).

With its emphasis on mindfulness, responsibility, and resilience over fast mass consumption of people and places, slow food tourism to places animated by bioregionalist concerns aligns with the cosmopolitan interest in authenticity. It values preservation and conservation not only of the land but also of cultural heritage. Food experiences that reflect bioregionalist concerns for sustainability,

DOI: 10.4324/9781003455516-7

conservation, cultural heritage preservation, and commensality also meet the moral needs of cosmopolitan travelers.

However, bioregionalist foods and foodways are in short supply in US national parks. In general, according to Slocum and Curtis, "local foods and culinary experiences do not currently play an essential role in national park experiences" (Slocum and Curtis 2016, 153; see also Sharples 2003). Indeed, the Institute at the Golden Gate reports that on publicly protected lands, the "quality and type of food served does not contribute to a park's environmental mission or unique sense of place" (2011). This conclusion aligns with the sentiment expressed in many of the park restaurant reviews I analyzed in Chapter 5.

One common element of bioregionalist slow food tourism is the experience of niche food products at the site of production. Visitors engage actively with local providers, exchanging knowledge as global cosmopolitan citizens (Clancy 2017). Although US national parks are generally associated with very few iconic food products, national parks in other countries are built around thriving local food industries. In the UK, for instance, the Lake District is known for Kendal Mint Cake, Grasmere Gingerbread, and Cumberland sausage, while Exmoor is famous for cider, cheddar cheese, and cream teas (Sims 2009, 322). That said, there are a small number of spaces in and around US national parks where these kinds of bioregionalist food experiences exist and draw tourists, cosmopolitan or otherwise, and these experiences stand in contrast to the largely industrial concessions system described in earlier chapters.

Various drivers are shaping the future of food tourism; on the one hand, scientific advances, scarce resources, and the quantification of the self point toward food tourism as a quest for novelty and innovation; on the other, desires for exclusivity and authenticity, increasing affluence, and investments in the experience economy push forward a future food tourism aimed at cultivating distinction (Yeoman and McMahon-Beattie 2015, 26). Local food experiences feel prestigious to cosmopolitan travelers, who relate them to ego-enhancement and self-satisfaction (Munt 1994; Kim, Eves, and Scarles 2009; Sims 2009). Cosmopolitan travelers—those with higher income and education level—are most likely to perceive food for its symbolic value, its ability to supply a sense of taste, beyond satisfying simple hunger (Bourdieu 1984; Glanz et al. 1998; Hall and Sharples 2003; Yeoman et al. 2015).

In contrast to the kind of tourism wherein tourist attractions are simply "a set of mirrors to serve the narcissistic needs of dull egos" (MacCannell 2011, 168), bioregionalist enterprises may offer the opportunity for a different kind of experience: one that "does not threaten the moral integrity of tourists or their hosts" (MacCannell 2011, 187). They may draw attention to their seams, to their very constructedness, and to the true richness of human difference rather than to a trivialized tourist-safe version.

Culinary tourism is a vehicle for the authentication of direct experience through culinary consumption; it tends to succeed when a cuisine's "exceptional quality and particular characteristics" can be extolled (Prentice 2001, 16). US national parks, born out of Romantic dreams of self-making in communion with nature, find themselves in a pickle regarding culinary tourism. Unlike the grand lodges of the early twentieth century, designed as mythic and majestic attractions in their own right (Carr 2007), the industrial foodscape of contemporary park restaurants is not well set up to deliver this kind of authentication. Consumers of tourism use several criteria to determine authenticity: the level of direct contact with the culture, the number of other tourists present, the level of independence in travel, and the extent to which experiences conform to expectations about landscape, climate, and culture (Waller and Lea 1999 cited in Prentice 2001, 11).

Cosmopolitans are not simply trying to emulate people who are richer, more cultured, or better educated. They want to consume romantically, in ways that are "imaginative, remote from experience, visionary, and preferring grandeur or passion or irregular beauty" (Campbell 1987, 1). Romantic consumption becomes, in this conception, almost a form of production, a craft activity involving skill, expertise, passion, and judgment (Campbell 2005, 27), and resulting in the production of the self. The national parks themselves serve this function well, except for the eating part. With limited opportunities for the types of self-defining cooking trumpeted by foodies (de Solier 2013), parks are places where restaurant consumption is pretty much the only game in town (except for picnicking, which the NPS is gaga about, as I described earlier). There's a lot of moral pressure on cosmopolitans, heirs apparent to the Romantics, to be productive in their leisure. Culinary tourism that is experiential and aligns with bioregionalist priorities fits the bill, in the absence of cooking opportunities.

This kind of tourism evokes the authentic and in doing so appeals to cosmopolitans. Unlike the world of industrial park food, with its plastic cafeteria trays full of food that is mass produced elsewhere, frozen, and then shipped and reheated, the inconvenience of a foodscape that depends on place, season, and local knowledge is highly appealing to cosmopolitans. Bioregionalist food experiences provide opportunities for

> aspirational eating, a process in which people use their literal tastes—the kinds of foods they eat and the way they use and talk about food—to perform and embody a desirable class identity and distinguish themselves from the masses.
>
> *(Finn 2017, 11)*

In her work on foodies, de Solier describes a process of self-formation characterized by moralities of productive leisure and a strong belief in the immorality of distinction (de Solier 2013, 16). In the moral hierarchy of productive leisure, producing food is better than just consuming it. Bioregionalist food enterprises offer a way forward here, as they emphasize production and "class up" consumption by emphasizing commensality in the eating experience. The concessioners, National Park Service (NPS), regional nonprofits, and private offsite businesses function as interpretive intermediaries as they emphasize productive leisure through tastings, food trails, and the like.

Bioregionalist Food Tourism in US National Parks

Surveying the foodscape in and around today's parks, I see three different types of existing bioregionalist food tourism initiatives. First, as described in more detail in an earlier chapter in the case of Yellowstone, I see industrial concessioners making bioregionalist gestures, for instance, by indicating local sourcing on their menus. Second, I see park-adjacent private and nonprofit bioregionalist initiatives that typically take the form of regional culinary trails. And third, I see a handful of NPS-sponsored, in-park bioregionalist programs.

Gestures by Industrial Concessioners

I discussed some of the bioregionalist gestures of Xanterra in Yellowstone restaurants and in fancier restaurants like Aramark's Metate Room, Grand Teton Lodge Company's Jenny Lake Lodge Dining Room, and Dawnland/Ortega's Jordan Pond House in earlier chapters. For instance, Jordan Pond House offers some local food and beverages and a few regionally traditional foods, and its parent company touts a corporate sustainability pledge. These are all gestures toward bioregionalism that appeal to cosmopolitan tourists, who are targeted with campaigns like the National Park Foundation's "4 National Parks for the Foodie in You," which acknowledges that "A thirst for adventure and a taste for the finer things go hand in hand at many of our national parks" (National Park Foundation 2016).

Another industrial concession making bioregionalist gestures is found at Shenandoah National Park, where the corporate concessioner Delaware North collaborates with local beverage producers. Their website claims,

> Farm-to-table offerings are the heart of Shenandoah National Park's culinary scene. That's why our culinary team pairs events with local wineries, distilleries, orchards, and breweries so you'll not only experience local beverages – you'll get to interact with the artisans that craft them.
>
> *(Delaware North 2022)*

They offer a Skyland Whisky-Wine Shuttle to local distilleries and vineyards, as well as local wine, craft beer, and hard cider tastings on-site at lodges in the national park and the occasional chef cooking demo. If cleanliness is next to godliness, perhaps interacting with and being educated by the artisans who produce the food has morally productive allure; there's also some degree of commensality to boot. Delaware North promises an authentic culinary experience connected to place.

The bioregionalist gestures made by industrial concessioners in the parks are certainly worthwhile and appreciated by visitors and the local producers and growers from whom products and ingredients are sourced. Some of these gestures are even well promoted, with bigger players like Xanterra, Aramark, and Ortega using publicity to draw attention to their efforts. However, these gestures do not yet transform the essentially extractive character of industrial concessions, nor do they harness the purchasing power of the NPS or its concessioners in any significant way; the setup is far more boutique and limited and seems to have frustrated chefs who are hoping for more sustainable local and regional sourcing than their parent company allows (Repanshek 2009).

Culinary Trails and National Parks

The next type of bioregionalist food tourism associated with national parks is park-adjacent, rather than within or run by the parks, and comprises primarily private enterprises. Most of these are culinary trails of some type, which are cooperative marketing mechanisms associated with alternative food networks that play to the identity of the cosmopolitan culinary tourist (Mason and O'Mahony 2011). These park-adjacent culinary trails emphasize local heritage cuisines as a means of branding destinations and realizing regional economic development (Hashimoto and Telfer 2015).

One example of a park-adjacent culinary trail is the Olympic Culinary Loop (OCL), an organized consortium of growers, producers, and restaurants that markets to culinary tourists exploring the area in Washington State that surrounds Olympic National Park. Their promotional materials—a locally distributed brochure and a website, OlympicCulinaryLoop.com—underscore Native American heritage, sustainably grown and locally harvested foods, handcrafted wines, and farm-to-table experiences, all catnip to cosmopolitans. Whether touting the annual Brinnon ShrimpFest and summer concerts at Olympic Cellars Winery, offering a printable map of "tasty itineraries and delicious destinations," or reporting on the unique flavor profiles of entries in an Olympic Peninsula Seafood Chowder cooking contest, the OCL drives home the idea that the peninsula is a destination for food that is fresh, local, seasonal, and sustainable, thus justifying more than $263 million dollars spent annually in nearby communities by visitors to Olympic National Park (National Park Service 2015).

Another example of a park-adjacent culinary trail is the Redwood Coast Culinary Tour, a self-guided experience promoted by the Humboldt (CA) County Visitors Bureau. Their Google map is not exactly slick, but their website sounds all the right notes for the cosmopolitan visitor to California's Redwood State and National Parks, who is encouraged to do a bay oyster tour by motorboat, make a visit to the country's first United States Department of Agriculture-certified organic brewery, and shop at the Arcata Farmers' market, "a bustling kaleidoscope of beautiful fresh flowers, colorful locally grown produce, and funky artisan goods created by North Coast crafts persons" (Humboldt County Visitors Bureau 2022a). Practical experience and exclusive access are highlighted, as the website implores visitors to participate in "hands-on oyster harvesting," a behind-the-scenes brewery tour, and a well-informed stroll through the historic Main Street District of Ferndale, "where art galleries, antique shops and boutiques complement a dazzling array of 19th-century Gothic Revival, Italianate, Eastlake and Queen Anne homes" (Humboldt County Visitors Bureau 2022b). Here, a tourist's capacity for evaluating features and making distinctions can be deployed. These opportunities, combined with the obvious charms of the parks themselves, have resulted in massive visitor spending: Visitation to Redwood National Park, combined with visitation to the three California State Parks within the Redwood National and State Parks partnership—Del Norte Coast Redwoods, Jedediah Smith Redwoods, and Prairie Creek Redwoods State Parks—results in over $90 million in annual community spending (National Park Service 2016a).

These park-adjacent bioregionalist culinary trails represent a familiar culinary tourism imaginary and direct revenue into the hands of small business owners, many of which operate in sustainable ways. However, there is little regulation or oversight to ensure that bioregionalist standards are met.

Food in the Parks: Spaces of Consumption and Production

Apart from these park-adjacent culinary trails that highlight bioregional principles appealing to cosmopolitans, there is a third type of bioregional food tourism related to national parks, and this kind happens within the parks themselves. Its forms, however, are varied. In some cases, rural environments and landscapes are conserved in national parks through the integration of tourism, leisure activities, and agriculture, increasing economic value to farming communities (Morris and Romeril 1986). "Consumption" countrysides have been returned, at least in part, to "production" countrysides, countering trends that see farmers, pressured for survival, into diversifying their attractive landscapes into spaces for tourist consumption (Pfeiffer et al. 2009). Additionally, some in-park bioregionalist projects exhibit themes of managing protected areas for the promotion of health and well-being (Thomsen, Powell, and Allen 2013).

The most well-known example of bioregional food tourism in a US national park involves the orchards of Utah's Capitol Reef National Park, where many tourists go precisely to harvest their own fruit and nuts from heirloom trees that have been maintained using heritage techniques. The u-pick orchards show up frequently on lists of the top things to do in the park (Park Ranger John 2022; Sleeping Rainbow Adventures 2022), and they draw tourists who get not only authenticity but the bragging rights of doing some of the morally precious labor associated with food production. Maintained by a two-person crew and aided by community volunteers for special rehabilitation projects, the Fruita Orchard compels reflections like from this pick-your-own visitor:

My own paper sack bursts with apples already, and it becomes apparent how important protecting this sanctuary in the Utah desert really is. I could have easily driven a few miles to my local supermarket to buy commercial apples, but I wanted to experience this culturally significant haven for myself. The physical act of plucking apples straight from the tree serves as a connection to not only the land, but also the rich cultural practices that have shaped this area for centuries.

(Nalewicki 2021)

Gathering apples from the Fruita Orchard is, for pickers like Jennifer Nalewicki, more meaningful than simply filling a bag at the grocery store and for all the reasons that cosmopolitans seek: It offers a place-based experience that is understood to involve culturally important labor.

Another less well-known bioregional food tourism in the parks initiative is Farms in the Parks (known as "Countryside") in Ohio's Cuyahoga Valley National Park (CVNP). The park partners with a nonprofit conservancy to preserve rural parklands by bringing historic farmsteads back to life. Farmers must use sustainable practices and be tourist-friendly, and the initiative is applauded for bringing over 100,000 people into the parks each year, increasing food access for local communities, and putting money into the pockets of small business owners (Countryside 2022). CVNP is a heavily visited national park, given its proximity to populous Cleveland and Akron, and locals know about Countryside: The farmers market is covered substantially in local media (Bigley 2016; Conn 2022).

There is a good website devoted to this initiative (CountrysideFoodandFarms. org), but the fact of its existence is not well promoted to park-goers; aside from a passing mention in the official park brochure, devoid of crucial details like where and when the farmers markets are held, one must scour the official NPS website for the park to know anything about Countryside. There's no information in the "Food and Dining" section of the CVNP website pertaining to Countryside, but five layers deep under "Learn About the Park," information is available, including audio clips from interviews with leaders involved in the Countryside project (National Park Service 2022d). In one of them, somewhat ironically given how deeply buried this content is, Local Food Programs Coordinator/Markets Manager Beth Knorr states that the high degree of visibility of the national parks provides a "bully pulpit to talk about the importance of sustainable agriculture" and shows that agriculture doesn't have to be destructive (Knorr 2021). The ways in which Countryside reconnects food to local time and place appeals to eaters who seek interaction with producers; this return to preindustrial ways allows people to express their resistance to postindustrial global modernity (de Solier 2013, 103).

The bioregionalist efforts of Countryside and its peers are not well known outside of local communities. Similar innovative sustainable agriculture projects exist in other NPS units, like the Martin Van Buren National Historic Site (MVB NHS) in Kinderhook, New York. Here the NPS collaborates with Roxbury Farm to preserve President Van Buren's historic farmland by supporting sustainable agriculture, in recognition of Van Buren's insistence that "progressive agriculture techniques could increase profitability and sustain future generations of farmers using free labor rather than enslaved individuals" (National Park Service 2020).

Roxbury Farm is a 400-acre family farm that provides community-supported agriculture (CSA), a partnership between the farm and community supporters who buy subscriptions; 20 acres of its land and some of its buildings are leased from the NPS (Roxbury Farm CSA 2022). When I asked MVB NHS Park Ranger Zach Anderson during a site visit, he acknowledged that locals know about the CSA, as it has nearly a thousand members, but that with no real public-facing profile or any recognition in the MVB NHS promotional material, the relationship of Roxbury Farm to the NPS amounted to a well-kept secret.

The last in-park bioregional food tourism program worthy of mention here is a little different, in that it is based in a National Recreation Area (Golden Gate NRA in California), rather than a full-on park like Capitol Reef or Cuyahoga Valley. Like in the Cuyahoga Valley model, Golden Gate NRA partners with a conservancy, and their work product is (among other things) a Food for the Parks initiative that aims to expand the availability of nutritious, local, and sustainable fresh food to park visitors nationwide. They want to leverage the buying power of the NPS and its supply chain to affect food systems. They have attempted to do this by creating case studies and a tool kit of sustainable food "best practices" case studies that parks and concessioners can adopt (Golden Gate National Parks Conservancy 2022). The entire enterprise is centered around increasing health, with "childhood obesity" and "rising healthcare costs" some of the chief villains to be vanquished (Institute at the Golden Gate 2012). The impact of this program is difficult to gauge; NPS Director Jon Jarvis convened a Healthy Parks, Healthy People US meeting at Golden Gate NRA in 2011, and it was attended by over 100 leaders in various industry sectors as well as NPS staff, but the extent to which the ideas discussed were then put into play in actual parks is unclear. The same lack of information exists about the other work products of the Golden Gate National Parks Conservancy—the case studies and tool kit—which are available on the site but put to unknown uses.

These various models of bioregionalist food experiences in the parks stand to be transformative, as they are carefully managed by the NPS and integrate thoughtful bioregionalist standards. However, they are not yet exactly well promoted, but there is great potential. The travel dollars spent to embody and exemplify cosmopolitan discourses involve money that can be used to support more socially just, sustainable, and environmentally responsible food systems, and I am eager to witness this work as it unfolds in and around US national parks. The kinds of bioregionalist food tourism initiatives I have described in this chapter attempt to "territorialize," to situate communities, consumption, and production in the local (de la Barre and Brouder 2013, 215). Despite many challenges including inaccessibility, high costs, and seasonality, bioregional food-focused tourism development in the national parks poses an intriguing path to ecological wellness.

The Institute at the Golden Gate has identified several challenges to bioregionalist food efforts in US national parks: keeping customer wait times short, maintaining appropriate facilities for food preparation, responding to customer food preferences, managing supply and sourcing issues, keeping prices comparable to other local options, and even complying with NPS regulations which, for instance, limit the sourcing of produce from non-USDA-inspected sources like farmers markets (2011, 5). Still, a concessioner's environmental protection efforts are favorably evaluated per the Concessions Management Improvement Act of 1998, so companies wishing to do business in the parks are incentivized to make gestures toward bioregionalism.

Indigenous Foodways, Culinary Tourism, and the Parks

In looking at all these different models that indicate modest successes for bioregionalist food tourism, I notice that the original local foodways—those of Indigenous people—remain largely obscured in national park spaces, rarely referenced by the industrial concessioners that hold sway over most of the eating in the parks. In this section, I analyze, evaluate, and propose some creative ways that food is being reimagined in the parks by Indigenous people and their supporters in the NPS and corporate concessions. My question for future national park food ecologies is: What would food experiences in the parks look like if they respected Indigenous food practices without appropriating them (Yazzie 2019), supported local producers, and made use of local ecological knowledge for the bioregion (Stevens 2020)?

The food scene in US national parks today is dominated by highly industrialized concessions operated by large corporations. In the previous section, I explored some of the challenges this configuration presents to environmental sustainability and discussed initiatives in- and outside the parks that attempt to improve sustainability. Tied to this set of problems, the industrialized character of US national park concessions also presents challenges to the livelihood of Indigenous people. But as is the case with environmental sustainability, several creative food-focused projects and initiatives exist that are aimed at not just at countering the imperilment of the original locals, but at celebrating and expressing Indigenous culture in meaningful and even profound ways. These projects and experiences align with cosmopolitan traveler preferences for experience-based culinary tourism and a careful consideration of their affordances and constraints points toward a palatable future ecology for food within the parks.

One does not usually think of national parks as culinary tourism spaces, but putting this frame around them allows us to think critically about how the relations among land, culture, identity, and foodways are articulated in the parks. Culinary tourism is "is a space of contact and encounter, negotiation and transaction" (Kirshenblatt-Gimblett 2004, xi–xii) between producers and consumers who enact social roles, express identity, and satisfy aesthetic needs in the process (Long 2004, 12). The desires of contemporary gastronomic tourists, always searching for authenticity in the spaces of the Other—different times and cultures, simpler lifestyles, cuisines "uncontaminated by the very market forces that enabled them to experience the authentic" (Scarpato and Daniele 2003, 300), tend to be centered in these encounters. But although consuming local food products on their park trips may cast an annoying halo of responsible/sustainable onto self-satisfied tourists (Sims 2009), is that alone reason to turn away from Indigenous and local foodways? Rather than being the "visited Other," Indigenous people, given their long connections to the lands that national parks now occupy, can be leaders in tourism; tourism studies scholars note that Indigenous people's historical commitment to sustainable development can manifest in unique tourism approaches that substantially benefit their home communities (Carr, Ruhanen, and Witford 2016, 1075). However, given longstanding power asymmetries and a history of government-sponsored violence toward American Indians, one must approach any consideration of the possibilities and perils of Indigenous culinary tourism in the parks with caution.

Hinch and Butler (1996) assess examples of Indigenous tourism based on whether cultural themes are present or absent, and whether Indigenous people control the visitor experience. They developed a heuristic for Indigenous tourism involving four types: *culture dispossessed tourism* (cultural themes; low Indigenous control); *Indigenous culture-controlled tourism* (cultural themes; high Indigenous control); *diversified Indigenous tourism* (no cultural themes absent; high Indigenous control), and *non-Indigenous tourism* (no cultural themes; low Indigenous control). This framework adds an additional dimension to my analysis of food-related offerings and initiatives in national park restaurants, and I incorporate references to it in what follows. However, I need to be clear that Indigenous people should define their own interest in involvement in national park food, and that the mere absence of Indigenous themes or control is not necessarily a sign of trouble, because maybe that's the way Indigenous people want it.

Indeed, most culinary tourism in the national parks can be characterized as non-Indigenous, as there are no cultural themes present and little to no control of the enterprise by Indigenous people. However, we do see cultural themes present and/or Indigenous control in several gestures by concessioners, private or nonprofit enterprises near parks, and even in a handful of initiatives within the parks themselves.

Co-management

National parks affect the well-being of Indigenous people in many ways: Issues include livelihood security, cultural and spiritual integrity, psychological well-being, governance, and educational and economic opportunities in park management and tourism (Scherl 2005). Indigenous people's involvement in cultural interpretation and tourism is an important element of the evolving relationship between the US federal government and Indigenous people in national park spaces (Zeppel 2009, 265). Scherl and Edwards conclude that community ownership and operation of tourism enterprises in protected areas provides Indigenous people living in or nearby with the strong opportunities for poverty alleviation and sustainable development, as one element of a diversified economy (i.e., not the only game in town, given how fragile the industry is) (2007, 81). "A rigidly dichotomous view of cultural tourism, in which 'guests' are sophisticated affluent members of powerful industrialized western societies and 'hosts' are naïve poverty-stricken members of powerless underdeveloped nonwestern societies, is clearly outmoded" (Guyette and White 2003, 165).

What if Indigenous people were engaged in the co-management of the national parks and the food experiences offered in them? This solution stops short of the wholesale return of the parks to Native Americans advocated by David Treuer in a prominent 2021 piece in *The Atlantic*; Treuer notes that without a strong tax base or significant non-extractive commerce, Native Americans remain dependent on federal support, and he offers up interpretive management of heritage and culture as an alternative (Treuer 2021, 40). He suggests transfers of control in Australia and New Zealand to Indigenous people as examples. Treuer doesn't get into the specifics of food or concession management, however. It would be wise for anyone genuinely contemplating advocacy of co-management to operate with what Tuck and Yang (2012) have described as an "ethic of incommensurability"—to recognize the reality that the future of settler colonial societies, even when making gestures toward the empowerment of Indigenous people, is fundamentally at odds with the true sovereignty that would result from a return of the land. Co-management needs to be real, rather than an exercise, a "settler move to innocence" (Tuck and Yang 2012, 9), to expiate white colonial guilt.

Co-management of US national parks dates to the 1980s, with the dawning recognition that national parklands had not historically been uninhabited wilderness, but rather sites of Indigenous cultures; the federal government, through laws like the Alaska Native Claim Settlement Act of 1971 (ANSCA) and the Alaska National Interest Lands Conservation Act of 1980 (ANILCA), began to embrace a participatory model of conservation, rather than a protectionist one (Zeppel 2009, 269; see also Catton 1997). However, although this model addresses subsistence use, including hunting, gathering, and fishing, and possibly some employment in government agencies, there are fewer agreements that pertain to tourism and interpretation (Zeppel 2009, 276). Some downplay the importance and impact of interpretation in a land-focused bureau like the NPS, suggesting that being authorized to talk about the land is different from having the power to manage it (Burnham 2013, 260). This perspective, however, gives short shrift to the power of narrative to shape discourse.

National parks, like museums, face questions about the politics of interpretation as they negotiate the conservation of heritage. "Who are the owners/custodians of the areas? How are they and the areas represented? Who speaks for them? What is spoken and why? Who is listening to the messages?" (Staiff, Bushell, and Kennedy 2002, 97). US national parks are increasingly incorporating Indigenous understandings of the landscape as well as teaching about Indigenous dispossession as part of visitor center displays and ranger talks, but there is relatively little that concerns foodways, and the concessioners authorized to provide food seem to have a very limited interpretive charge. But there is a case to be made for the inclusion of other sensory perceptions, besides Eurocentric visuality/ocularity, in the parks—an emphasis on taste would deepen the experience of these spaces and allow for the expression of marginalized foodways.

"Education, activism, and policymaking must support Indigenous Peoples as the narrators of their own gastronomies, as well as their determinative say in whether and how they will participate in the national culinary culture" (Grey and Newman 2018, 726). This seems like reasonable direction for Indigenous gastronomy in national park spaces and aligns with guidance to center host populations in the development of cultural tourism opportunities (Guyette and White 2003).

Lest concession management by Indigenous people seem like an unambiguous win, one must note the critiques of this kind of model. Critics deride co-management as "a state-ratified international rights regime" that "cannot do other than undermine Indigenous self-determination and imperil Indigenous peoples' cultural heritage" (Grey and Kuokkanen 2020, 919). They suggest, instead, a model of Indigenous governance, which allows cultural heritage to thrive by being actively experienced in the original place. An excellent direction for future research would involve deeper exploration of structural transformation of park food experiences that fully integrate Indigenous governance, enabling Indigenous self-determination, and protecting and celebrating Indigenous people's cultural heritage.

Gestures by Concessioners: Décor, Dishes, and Displays

Contemporary Indigenous foodways are largely overlooked in US national parks today. A 2022 National Park Foundation web article about NPS units that celebrate Native American heritage mentions the ancestral uses of various parklands for farming and gathering foods (Watson 2022), and visitor center exhibits throughout the national parks often feature panels describing precontact Indigenous food production practices. It is also not unusual to hear park rangers describe historic Indigenous food practices as part of their ranger talks, especially in parks like Mesa Verde that center Indigenous histories. This focus on past foodways has the unfortunate effect of freezing American Indians as "an idea and an artifact, a static and quaint people who have few economic needs" (Keller and Turek 1998, 178).

Food is marginalized as a form through which NPS units communicate about *present-tense* Native American culture. However, there are instances in which large corporate concessioners are trying to incorporate Indigenous cultural themes into food experiences. Large industrial concessioners occasionally reference Indigenous culture through the décor of their restaurants, particularly the more upscale ones. Mesa Verde National Park's Metate Room, discussed in an earlier chapter, is one of the most notable, given the Ancestral Puebloan cultural focus of the park. The Metate Room website boasts in 2022 that "Colorful Native American artwork surrounds you as you enjoy our contemporary menu inspired by regional heritage foods and flavorings," and the walls of the restaurant are adorned with striking photographs, attractive Navajo and Ganado textiles, brightly colored paintings of animals, and a display of woven baskets. It is not clear who created the artwork, whether they have Native American heritage or how they are compensated, but the décor functions to create a vaguely "southwest" vibe meant to accompany the restaurant's "contemporary, sustainable interpretations of heritage foods with a Southwestern flair" (Visit Durango 2022).[1]

Similarly, another prominent visual reference to Native Americans in a park restaurant is found in Grand Teton National Park's Mural Room, where the high-end restaurant at the Jackson Lake Lodge is named after the resident artwork. Created by New York-based settler artist Carl Roters, the ten-panel "Rendezvous Murals" depict an 1837 meeting of fur traders. It's all tipis and feathers, bows and arrows, and jovial and joyous cooperation among trappers and traders, with little sense imparted of the damage that settler fur trading would do to Indigenous food resources. In these cases, despite the inclusion of Native American motifs in the décor, the lack of involvement of Indigenous people in the creation and curation of the art would lead Hinch and Butler to suggest that these gestures exemplify "culture dispossessed" tourism (1996). The Native American visual motifs are a relatively superficial part of the *mise-en-scène* of restaurants run by large industrial concessioners.

Beyond décor, corporate concessioners sometimes gesture to Indigenous foodways in their menus and in the way they talk about the food they serve. At Metate Room, Aramark touts a menu "inspired by regional heritage foods and flavorings," which suggests Indigeneity, but there is scant evidence of foods or flavorings typical of Ancestral Puebloan cultures on the menu. In the summer 2021 menu, there was no indication whatsoever about product sourcing and nothing to suggest preparations that draw on Indigenous or even non-Indigenous regional foodways, just the typical fish-beef-chicken-pasta assortment of entrees that one finds at almost any upscale park restaurant.

In Utah's Zion National Park, Xanterra's Red Rock Grill offers, among its lunch entrees, an "Authentic Navajo Taco," involving all the usual taco fillings stuffed inside Navajo fry bread. (The complicated circumstances under which fry bread, a product of colonialism, has come to be endorsed as "authentic Navajo" warrant their own entire treatment, but that is not my focus here.) There is a note on the menu indicating which items meet Xanterra's sustainability standards for local, third-party certified, and animal welfare, but there is no information about producers or sourcing, so it is difficult to ascertain connections to Indigenous foodways. However, three sisters foods (corn, beans, and squash) are liberally utilized throughout the menu, suggesting at least some degree of interface with traditional Indigenous foods.

Xanterra makes its commitments to Indigenous foods more explicit at its Arizona Steakhouse in Grand Canyon National Park, where its website informs readers that the

> newly enhanced dining experience tells the epicurean story about the native ingredients, flavors, and history of Arizona and its people with an emphasis

on heritage/heirloom ingredients ranging from Green Chile to Tepary Beans. These indigenous foods are more colorful and more nutritious. Whenever possible, these specialty items are sourced from Arizona family farms and ranches and Native American businesses. Supporting small farms and artisan producers is good for Arizona, and using their products is good for the consumer. Our culinary team combines these ingredients with the freshest and most sustainable meats, fish, and produce available, creating a delicious, educational, and memorable dining experience.

(Xanterra 2022)

The menu incorporates chiles, tortillas, cotija, corn, beans, and prickly pear, in a gesture that does in fact have some relationship to Indigenous foodways. In this way, we see corporate concessioners making interpretive gestures by linking the food they serve to both Indigenous traditions and the local lands.

One might argue, however, that the specificity of ancestral Indigenous foodways has been lost, in favor of the same kind of commingled regional style that characterizes NPS architecture ("parkitecture"). As Rothman notes, early park concessioners extended industrial capitalism's reach, providing services that eased the burdens of travel for a public disinterested in distinguishing among varieties of mythic Indians (Rothman 1998, 70). In this context, any concessioner gesture toward Indigenous foodways seems beneficent and responsible, even if the foods and preparations featured in concessioner-run park restaurants are generic, with a relatively tenuous connection to the specific local culture. It is easy to finger wag at concessioners for missing the mark, but they are serving a market: If hungry tourists don't care about the particulars of whether their park food experience reflects and honors the ways that Indigenous people used the specific lands, nothing changes. This is precisely why the demands of cosmopolitan travel, however fraught the quest for "authenticity" is, intrigue me—they power some worthwhile developments. I'll expand on this in the next section.

There is room for improvement in concessioner gestures that focus on generic dishes or décor. The presentation of Native American motifs in park restaurant décor tends to preserve a highly romanticized view of precontact American Indians. And foodwise, how meaningful is the inclusion of corn and beans on the menu, if those products arrive in a giant Sysco truck? Where is the décor in the restaurants and the sort-of-Indigenous food coming from, anyway? Could the concessioners buy directly from American Indians instead of ignoring their food altogether or imitating their food poorly? And what are concessioners, whose customers may be attracted by the inclusion of Native American themes, doing to share their profits with Indigenous people? Are they, as suggested by Dave Simon of the National Parks and Conservation Association, "donating a portion of their waived fees to fund ethnographic studies, improve interpretation of Indian subjects, and protect sacred sites" (quoted in Burnham 2013, 281)?

An organization called the American Indian Alaska Native Tourism Association (AIANTA) is working to build alliances between Indigenous people, concessioners, and the national parks that benefit Indigenous people. AIANTA employs agritourism specialists and partner program coordinators who are focused on making these connections, allowing more thoughtful collaborations between concessioners and Tribes to emerge. AIANTA helps concessioners to recognize their role as interpretive intermediaries whose embrace of American Indian cuisine can have a positive economic and cultural impact on local Tribes. A partnership in Mount Rainier National Park got underway back in 2019, when Rainier Guest Services, which operates the Paradise Inn Dining Room, as well as the National Park Inn Dining Room, sourced smoked salmon directly from the Nisqually Tribe, working it into a popular smoked salmon dip served with green apples, showcasing some of classic flavors of Washington State and the Pacific Northwest. The culinary team at these restaurants also worked to develop specials that were inspired by traditional foods. Brandy Frederich, Senior Director of West Coast Operations for Guest Services, which operates concessions in the park, explained in an interview that the Nisqually Tribe's smoked salmon enterprise has subsequently suffered due to climate change, and they are no longer able to produce enough to sell to the concessioner.

In the aftermath of the COVID pandemic, Frederich and Rainier Guest Services have been working together with the park Superintendent, members of the affiliated Tribes (Cowlitz, Squaxin Island, Nisqually, Confederated Tribes and Bands of the Yakama, Muckleshoot, Puyallup, and Coast Salish Peoples), AIANTA, and other staff to develop opportunities and action steps related to cultural tourism in Mount Rainier National Park. Yakama Nation is collaborating with Rainier Guest Services toward the goal of providing visitors with locally sourced food through the concessioner. However, Frederich recognizes that the ability of some Tribes to produce food for such partnerships is challenged by climate change and other resource issues, and that concessioners need to do their homework to locate and develop trust and sustainable relationships with Native American food enterprises that might lack promotional resources.

Frederich is eager to help boost the signal of the local Tribes but is rightfully cautious about speaking on their behalf. For the time being, she is doing her homework, learning about the communities, foods, and arts of the Tribal cultures on whose ancestral homelands Mount Rainier National Park sits. Frederich reflects with gratitude on her participation in the 2023 Water Sounds Native art auction and traditional dinner sponsored by the Squaxin Island Tribal Council, noting that her experience of the impressive Squaxin culinary arts and handicrafts inspired her to enhance Rainier Guest Services' stewardship and interpretive messaging. She intends to work toward the goal of offering internships and hiring Tribe members to work with the culinary team to offer more traditional dishes and expanding collaborative business opportunities that push forward the interpretation of American Indian cuisine in the park.

Possibilities for this kind of more systemic change exist also within Eastern National, a nonprofit organization that "promotes the public's understanding and support of America's national parks and other public trust partners by providing quality educational experiences, products, and services" (Eastern National 2023). Among its many functions, Eastern National employs buyers and product developers who determine what gets sold at many national park stores; they also manage interpretive services at many park sites. They invite local and regional artisans to submit product proposals in a variety of categories, including food, and Eastern National claims that each item they sell goes through a stringent evaluation by NPS staff for quality, historical accuracy, and relation to the site's theme. There is little evidence in their Annual Reports that this has produced any momentous opportunities for Indigenous food in the parks, but the possibilities are suggestive, particularly given their existing ventures that recognize the interpretive function of culinary products—for instance, they sell a Mary Todd Lincoln cake mix at the Lincoln Home National Historic Site. Eastern National's "Mission 24 Strategic Plan" calls to "engage with vendors on supplier diversity and sustainability" and to "provide direct support to public-lands partners for expanded and more inclusive storytelling" (Eastern National 2023). The emergence of these types of commitments opens the door to more substantive or systemic engagement with Indigenous foodways in the parks.

Fortunately, these modest and emergent concessioner gestures toward Indigenous foodways are not the only game in town. Other ventures—private, sometimes nonprofit, food enterprises near national parks that benefit Indigenous people—demonstrate that food and commensality have a role to play in the disruption of white people's romanticized views of old-timey American Indians and, more importantly, in Indigenous sovereignty. Looking closely at a few examples, I assess whether commensal foodways-centered interactions might "lead tourists to see themselves as visitors in Indian country and not simply as pilgrims at an American shrine" (Spence 1999, 139).

Private/Nonprofit Enterprises near/with Parks

Despite justifiable critiques of cosmopolitan priorities and concerns, the needs and desires of cosmopolitan travelers often align with and support private and nonprofit food enterprises that center Indigenous cultures. Cosmopolitan travelers are big proponents of the experience economy (Pine and Gilmore 1999), wherein goods and services are differentiated not by their inherent value but by how they are consumed. Providing meaningful experiences to consumers attaches significant value, and American Indian entrepreneurs have realized that allowing tourists to witness or even participate in food production and communal consumption is a way of promoting local products and Indian-owned businesses. This kind of food tourism gives cosmopolitans "the pleasures of material culture without the evidence—or guilt—of accumulation" (de Solier 2013, 81). It also helps to ease cosmopolitan anxiety about simply consuming all the time, and it also helps them distinguish between different places and cultures, correcting for the generic Indian-ness that is presented in many industrially run park concessions. The commensal experiences these enterprises provide assure that "they are witnessing a form of counter-globalisation in which mass, anonymous factory production is replaced with something more local, individual and human in scale" (Meethan 2015, 121). Although critics deride as racist the reality that tourists will spend more when they get to witness or interact with the Indian "Other" creating a product (Beard-Moose 2011), the practice does ensure that local creators are supported (Lewis 2019, 76–77).

One of the most interesting examples of this kind of enterprise is an Airbnb Experience hosted by Mariah Gladstone. Gladstone, a cooking show host, environmental scientist, and "Indigenous Cuisinologist" who has Blackfeet and Cherokee heritage, runs IndigiKitchen, an operation founded to celebrate traditional Native foods. The organization offers cooking videos, speaking engagements, and school programs, and has "branched out to supporting Native producers, building Indigenous gardens, and teaching some of the important background surrounding the colonization of our foodways" (IndigiKitchen n.d.). Gladstone is deeply involved with Indigenous food sovereignty efforts in and around her community in Northwest Montana in lands now occupied by Glacier National Park and has many balls in the air. One that connects to tourism and eating in national parks is her Airbnb experience.

Gladstone hosts a "Dinner featuring local Blackfeet foods" on the Blackfeet Indian Reservation in Babb, Montana, the gateway to the Many Glacier Area of Glacier National Park. The experience features a four-course meal that highlights the area's ancestral foods and consists of locally harvested ingredients (Airbnb 2022). Offered at "a secluded home on the shore of Duck Lake, with a view of Yellow and Chief Mountains," Gladstone asserts on her Airbnb Experience page that "this land has nourished my ancestors for over 14,000 years and the menu for the evening will consist of foods harvested in the surrounding landscape" (Airbnb 2022). Reviewers express positive sentiments about the quality of the food, commenting on the tasty soup, greens, wild rice, smoked salmon, and bison. However, most say that the best part of the experience is sitting and talking with Gladstone and her partner about traditional Blackfeet gardening and cooking methods, as well as Blackfeet culture and history, and they enthusiastically recommend the experience as a must-do cultural immersion while visiting Glacier National Park.

This Airbnb experience centers Indigenous foodways and benefits Gladstone and, by extension, her community. The experience constitutes what Pratt (1992, 7) calls "transculturation," as Gladstone, a member of subordinated groups (Blackfeet and Cherokee), selects and invents forms of cultural expression using elements of cosmopolitanism embedded in the dominant culture. In the experience, she invites park tourists into her home and her garden, ostensible "back regions" that fulfill the cosmopolitan craving for authenticity, and teaches them about the land they're visiting, as well as Blackfeet culture and gastronomy. Gladstone controls what she reveals, offering entry into a back region that "is really entry into a front region that has been totally set up in advance for touristic visitation" (MacCannell 1973, 597). This experience generates revenue for Gladstone and is one of many "hustles" that help to support her while she carries out food-focused activist community organizing work on other fronts (Gladstone 2022).

At the time of this writing, another American Indian-owned, park-associated food enterprise has been the subject of much buzz in cosmopolitan spaces, including recent features in *The New Yorker* and *Esquire* magazines: Owamni, a Minneapolis restaurant named as "the most prominent example of Indigenous cuisine in the United States" (Kormann 2022) opened in Summer 2021 in the Central Mississippi Riverfront Regional Park, a regionally popular public park. Although Owamni is not an enterprise near a *national* park, I include it here because its stratospheric success as a park-situated restaurant doing political work on behalf of Indigenous people is instructive, and entirely relevant to the discussion of food enterprises near national parks.

By all accounts, dining at Owamni is anything but a generic experience of vaguely pleasant Indian ambiance. Touted as a "fully decolonized" restaurant, the eatery has been praised for the extent to the dining experience confronts its customers with political facts; *Esquire* magazine describes it as a space "that'll ask diners to discuss the decimation of Native American foodways as they feast on cedar-braised bison" (Nelson 2021). Owamni uses only Indigenous ingredients, and none of the familiar artifacts of culinary colonialism like chicken, pork, beef, dairy products, cane sugar, or wheat flour, inviting diners to reflect on the socioeconomic consequences of their daily food choices. A pinkish-orange neon wall hanging in the restaurant glows, "YOU ARE ON NATIVE LAND," and the servers wear colorful t-shirts with "#86colonialism"—a reference to restaurant lingo for dishes no longer available—emblazoned on the back. Chef/owner Sean Sherman concludes,

> Just taking the time to taste and to learn gives people an opportunity to think about the wrongs that have happened. It gives us an opportunity to talk about history from an Indigenous perspective and to acknowledge these atrocities that have been wiped off our history books. And it gives people an opportunity to realize a lot of issues are still alive out there.
>
> *(Sherman, quoted in Nelson 2021)*

Owamni is but one project of The Sioux Chef, "a loose confederation of chefs, ethnobotanists, food preservationists, adventurers, foragers, caterers, event planners, artists, musicians, food truckers, and food lovers committed to revitalizing Native American cuisine" (Sioux Chef n.d.). Owners Sean Sherman and Dana Thompson also run the nonprofit organization NATIFS, short for North American Traditional Indigenous Food Systems. NATIFS reestablishes Native foodways through its Indigenous Food Labs, which carry out research and development; Indigenous food identification, gathering, cultivation, and preparation; and a training center that prepares people to operate culinary businesses that center Native traditions and Indigenous foods (Sherman 2019; NATIFS n.d.).

What would a partnership look like between the NPS and The Sioux Chef, or NATIFS? Could it create forward-thinking foods and food experiences in the national parks, which are spaces of historic American Indian disenfranchisement, as it has done in a small but popular regional park in Minneapolis? As we consider the possibilities, we must tread carefully: Enterprises that center cosmopolitan travelers and purely instrumentalize local/Indigenous foodways and commensality are against the spirit of bioregionalism. Nonetheless, the efforts of Mariah Gladstone, IndigiKitchen, and the Sherman/Thompson team behind NATIFS, The Sioux Chef, and Owamni are demonstrating the revolutionary energy that comes from centering Indigenous foodways in private for- and nonprofit enterprises that harness the power of "park place."

Indeed, there is support for the idea of Native Nation-run small businesses as politically and socioeconomically meaningful responses to the economic subjugation caused by settler colonialism (Lewis 2019, 4). Courtney Lewis is an expert on Cherokee-owned small businesses in the Qualla Boundary, at the gateway to Great Smoky Mountains National Park, and she writes convincingly about what travelers seek there. But Lewis is not primarily interested in the tourists—rather, she understands a variety of motivations among American Indian small business owners and rejects notions of the Noble Entrepreneur who just wants to give back to the community (2019, 27). These business owners are well positioned to respond to the changing requirements of travelers.

Prior to the last 1960s–1970s Red Power movement, tourists preferred a non-specific version of Indianness, eschewing authentic connections to Indigenous people in favor of a nostalgic experience that would distract from their own complicity in oppression (Lewis 2019, 57–58). However, the gains of Red Power have shifted tourist interest to authenticity, so much so that they now critique what they perceive as inauthentic as tacky, not real, and old-timey (Lewis 2019, 58). For instance, Lewis describes the menu of Granny's, an American Indian-owned restaurant in Cherokee at the Great Smoky Mountains National Park gateway that also offers catering. Granny's doesn't offer many "Cherokee foods" in the restaurant but does offer them when requested for smaller catered events. The owner Teresa says,

> If [visitors] are looking for Cherokee food, they're all going to be disappointed because it's just beans and potatoes and corn bread like everybody else has eaten forever. The Cherokee food's very bland. Most of these people with this educated palate [are] not going to like it. Of course, if you grew up eating it, it's really, really good.
>
> *(quoted in Lewis 2019, 73)*

This anecdote suggests that cosmopolitan tastes and interests will need to be retuned: Cosmopolitans will have to move beyond the "photographs of beautifully prepared dishes, backyard gardens, braids of corn and onions, and Indigenous people wearing chef's jackets" (Mihesuah 2019, 309) popular on social media. They will need to recalibrate their tastes, appreciating the simple tastes and confronting their complicity in historic and contemporary American Indian realities of poverty, drug and alcohol abuse, and diet-related illnesses (Mihesuah 2019, 309; see also Diamond 1992; Joe and Young 1994; Barsh 1999), as well as the joys of specific cultural identity. As places of moral and cultural significance, national parks are promising sites for rethinking Indigenous foodways, and this important work must continue.

Park-Based Initiatives

The emergence in recent years of several in-park interpretive centers focused on Indigenous people suggests that the US federal government has evolved beyond earlier stages of park/Indian relations, including those of unilateral land appropriation; continued neglect of Tribal needs, treaties, and cultures; and aggressive Indian pursuit of Tribal interests, and has reached a point of commitment to cross-cultural integrity and cooperation (Keller and Turek 1998, 233). Since the late 1980s, with the establishment of a Native American Relationships Management Policy, the NPS has pledged to respect and promote Tribal cultures as part of the parks themselves. It has also acknowledged that it needs to look to other countries for models of how economic activities not normally carried out in US national parks might be accommodated here and to build mutually beneficial exchanges with American Indian Tribes and other Native American organizations (Rogers 2009, 9). More recently, it has released new guidance to strengthen Tribal co-stewardship of national parklands and waters (National Park Service 2022c).

The NPS is increasingly publicizing its commitment to collaborating with American Indians, Alaska Natives, and Native Hawaiians to preserve Native cultural heritage and celebrate Tribal cultures, particularly through supporting the development of Tribal tourism (National Park Service 2016b). For instance, the website and travel guide co-developed by the NPS and the American Indian Alaska Native Tourism Association highlights Native American experiences along historic Route 66, including information about authentic cultural experiences like pow-wows, festivals, and restaurants (AIANTA 2023). Although these collaborations have largely not focused on foodways, there is room for development in this area.

At present, the NPS stops short of promoting Indigenous foods, unlike its legislation-prompted treatment of Native arts and crafts. Rather than requiring concessioners to show how they are promoting Indigenous foods in a way that benefits Indigenous producers or offering incentives for such action, the NPS simply offers resources to concessioners to help them find local foods produced by Native Americans. They do this primarily by directing concessioners through a web page entitled "Procuring Local Foods" (National Park Service 2022a) to the website of the Intertribal Agriculture Council (IAC). Although great in theory, the site is not built out enough to be of much help, as it lists relatively few producers nationwide, and is as likely to list producers of moccasins or hand lotions as food products. The Council runs a variety of projects that could be impactful in national park food spaces, including initiatives to facilitate market access for Native food-related products and services, a "Made/Produced by American Indians" trademark to identify food products made by members of federally recognized Tribes and Alaskan Native Villages, and a retail partner program that allows businesses to identify themselves as offering foods made or produced by American Indians (Intertribal Agriculture Council n.d.). Through the work of the IAC, the tools for Indigenous food promotion in parks exist—they just need to be utilized.

In evaluating proposals from would-be concessioners, the NPS primarily looks for the concessioner's ability to protect and conserve park resources; their ability to provide the needed visitor services at reasonable rates; their prior experience and financial capabilities as a concessioner; and how much revenue the concessioner is likely to provide to the US government. Secondarily, the NPS examines the concessioner's environmental management programs and activities and its plans to employ or partner with American Indians (Concession Contracts 1998). These last considerations—environmental management and American Indian employment/partnerships—could be elevated to first-tier criteria for more meaningful and productive engagement in park foodways on both fronts. I do not mean to suggest that more American Indians should be employed in concessions to ensure that food is "authentic." To demand that the national and ethnoracial origin of the cook matches that of the recipe is to practice a problematic form of gastronativism that imagines an untainted and true, yet strangely generic, version of Indigenous gastronomy. Rather, I am interested in more strategic and consistent efforts in the parks to support Indigenous economies.

Inside some parks, one can find examples of private, nonprofit organizations that do important work around contemporary Indigenous foodways, even when it is not central to their mission. For instance, the Mesa Verde Association (MVA) runs a bookshop inside the visitor center at Mesa Verde National Park. On a June 2021 visit, the very first thing one saw when entering the shop was a display of Indigenous-authored cookbooks and Indigenous/locally produced foods. On offer were Bow and Arrow brand blue cornmeal, produced on the 7,700-acre farm run by the Ute Mountain Ute Tribe, as well as burlap sacks of Anasazi beans produced by the Adobe Milling Company in Dove Creek, CO, and cookbooks including The Pueblo Food Experience Cookbook, emerging from the Tewa Pueblo of Santa Clara. The MVA claims to subject all the products it sells to "a meticulous review process to ensure the greatest educational and interpretive value" (Mesa Verde Association Stores n.d.). It appears that this effort pays off, as the store benefits Indigenous people by buying their products and turns its profits back to the park, supporting interpretive programs, research activities, and visitor services. However, the number of food products is quite limited compared to Indigenous-produced handicrafts available in the park. This comparison might suggest a way forward.

Another promising development in centering Indigenous food cultures in the parks is found in Yellowstone National Park, where the Yellowstone Tribal Heritage Center (YTHC) opened in 2022. Prominently situated just steps from Old Faithful—the park's biggest draw—the YTHC emerged from a partnership of the NPS and Yellowstone Forever, a nonprofit educational and philanthropic organization aimed at benefiting the park and its visitors. Housed at the historic Haynes Photo Shop, the YTHC offered four months of programming involving visiting artists and presenters who interacted one-on-one with drop-in visitors, sharing with visitors about their Tribal heritage and their art form or area of expertise. Most of the presenters were artists or storytellers, but in 2022, the YTCH invited two presenters—Mariah Gladstone (Blackfeet) and Gwendolyn B. Carter (Nez Perce)—to educate visitors about traditional Indigenous foods. Rather than bemoaning the extent to which foodways seem peripheral to arts and crafts in these presentations of Tribal heritage, I argue that handicrafts offer an interesting model for future foodways work.

The Model of Handicrafts

A model might already exist for thinking through the recentering of Indigenous foodways in the parks. The National Parks Omnibus Management Act of 1998 ensures the promotion of the sale of Indian, Alaska Native, Native Samoan, and Native Hawaiian handicrafts (sec. 416) in US national parks. However, these regulations do not pertain to food. Although there is arguably a more complex regulatory framework around food production and consumption, the protections given to Indigenous handicrafts, including having their revenues exempted from concessioner franchise fees, could benefit Indigenous food producers and distributors. In 2022, the NPS convened several forums on retail sales of Native American handicrafts, inviting representatives from the NPS Commercial Services Program, the Department of the Interior Indian Arts and Craft Board, concessioners, and others to discuss how to ensure that products are "authentically Indian" and how to strengthen relationships with Indian artists and arts and crafts business owners (National Park Service 2022b). Jill Marsh, Corporate Associate Buyer from Xanterra, presented about the concessioner's efforts to find and feature Native American artisans local to the parks she buys for and to enrich the guest experience by connecting the travelers to artists and sharing origin stories for the art (Marsh and Trivelpiece 2022). It is not a stretch to imagine a similar approach to food.

The Indian Arts and Crafts Acts of 1935 and 1990 sought to grow the market for arts and crafts made by American Indians; to protect them from competition from inauthentic, mass-produced products; and to ensure that consumers were getting the real thing. Although food has not been historically protected in this legislation, there were many moments of recognition of food as a form of cultural expression in a 2017 US Senate hearing to modernize the Indian Arts and Crafts Act (Senate Hearing 115–75, 2017). Through the trademarking program of the IAC, which I discussed earlier, the mechanism exists for doing for food what has been done for arts and crafts. In national park spaces, the promotion of American Indian foods and foodways can not only satisfy the existential cravings of hungry tourists and fulfill the interpretive, educational mission of the parks but also honor Indigenous traditions while contributing to Indigenous economic prosperity.

As we move forward and see the development of initiatives to center and promote Indigenous foodways in the national parks, we must be careful not to fall into the kind of trap identified by Heldke, wherein locally produced becomes the shorthand for virtuous (Heldke 2012, 34). Just because a food product or experience embodies Indigenous cultural themes or is controlled by Indigenous people doesn't necessarily mean it is ideal. We also need to ask whether the product or experience has systemic impact or whether it is one-off and individualized. We need to center Indigenous people's interests and to acknowledge that sometimes that means that they are more interested in the restoration of their rights to hunt, gather, or fish on parkland than in running park concessions or doing interpretive work for the education of visitors. Indeed, this was the case with the Alliance to Protect Native Rights in National Parks, a group of six Tribes (Timbisha, Miccosukee, Pai 'Ohana, Hualapai, Navajo, and Sandoval Indian Pueblo) that banded together in the mid-1990s to press the NPS for unrestricted access to their ancestral lands (Wilkinson 1996), in just one example of the activist priorities of Indigenous people where parks are concerned.

The possibilities of Indigenous food tourism in national parks are manifold, but Nielsen and Wilson contend that Indigenous tourism "is still predominantly driven by the needs and priorities of non-Indigenous people" (2012, 67). Indigenous people must determine and manage their own engagement in tourism. Partnerships between the NPS and Tribes must emerge from the needs of Indigenous communities, transparently represent community goals and values, and ultimately, benefit those communities, rather than merely enhancing the vacations of the tourists visiting the parks.

Cautions about Tourism

Travel to national parks promises tourists an experience that conveys authenticity and that is culturally and environmentally sensitive (Munt 1994, 119). Of course, "authentic" is complicated. Food tourism that hinges on encounters with an "authentic" untouched by the flows of globalization or the reflexive fetishization of the local ignores its own constructedness (Parasecoli and Lima 2012). I have examined ways in which the food experiences in the parks fall short of this projected image and showcased efforts to increase cultural and environmental sensitivity. I'd like to offer a word of caution about psychically investing in tourism as a corrective to American Indian dispossession.

It is understandable to look askance upon promises of salvation through participation in tourist economies, given a plethora of historical examples where

Indigenous people got the short end of the stick in arrangements devised by the NPS or corporate concessioners. In the 1960s, the US Department of Commerce persuaded the Oglala Sioux Nation to authorize a Badlands land swap that, the government argued, would jump-start the local economy by showcasing Sioux culture for tourists; the NPS then awarded the Oglala Sioux the concession at Cedar Pass Lodge, which meant "dozens of (seasonal) jobs, a profitable outlet for Indian crafts, a chance for the tribe to run a major business in a tourist stronghold" (Burnham 2013, 142). Once the largest Indian-controlled concession in the NPS, with gross sales of $1.5 million and a staff of 80, most of whom were Native American (Burnham 2013, 234), the Cedar Pass Lodge also sold cheap tourist souvenirs over local handicrafts, replicating the missteps of the corporate concessioners. However, that concession is now run by Aramark, suggesting that Tribal control was relatively short-lived.

Corporate concessioners have been little better in their impact on Indigenous people. For example, one need only to recall the ways in which Blackfeet tribe members were employed in Glacier National Park in the 1920s and 1930s to promote a past-tense "romantic-savage-in- a-pristine-wilderness motif" (Keller and Turek 1998, 57):

> At resorts decorated with quasi-Indian art, tourists shook hands with Blackfeet 'chiefs' and flirted with 'Indian door girls.' The tribe drummed, danced, and signed postcards on the sprawling estate of a hotel, then conducted naming ceremonies in the crowded lobby. Indians 'passed the tom-tom' for tips and sold miniature bows, arrows, and tepees. Between 1914 and 1919, hardier tourists slept in real teepees on the lawn of East Glacier Hotel. When the popularity of such camping declined, the railroad hired Blackfeet to live on display in tepees.
> *(Keller and Turek 1998, 57)*

At the same time as they were entertained by the performative spectacle of imagined Indianness, tourists were protected by the NPS and Great Northern Railway, the concessioner, from the more troubling realities of contemporary Indian life. Blackfeet were enlisted to cater to the fantasies of white tourists but were not trained to participate in park management-run concessions (Keller and Turek 1998, 62). The same can be said of Yosemite Indians who performed at "Indian Field Days" in Yosemite National Park in the 1910s to goose up tourism. These field days consisted of a hodgepodge of Native cultural expression circumscribed by the expectations of whites—so the Yosemite were rewarded for dressing up in attire that would have been foreign to them but that jibed with what tourists expected of them (Spence 1999; Cothran 2010). And it was only lack of interest by Crow that kept concessioner E.C. Waters from putting them on display with some bison on an island in Lake Yellowstone in 1899, creating an "aboriginal exhibit" akin to the one at the recent World's Columbian Exposition in Chicago (Spence 1999, 69). Whether programmed or only dreamed of, the history of US national parks includes a regrettable number of problematic attempts to romanticize Indigenous existence for white audiences. Thus, I urge caution when thinking about how Indigenous tourism might be made productive for Indigenous people.

The tourist industry that has sprung up to serve the needs of cosmopolitans and other travelers in and around US national parks has become a vital economic driver, providing not only jobs but local revenue. However, as what Hal Rothman has identified as an extractive industry (1998, 13), it comes at a cost.

> Tourism is the most colonial of colonial economies, not because of the sheer physical difficulty or the pain or humiliation intrinsic in its labor but because of its psychic and social impact on people and their places. Tourism and the social structure it provides transform locals into people who look like themselves but who act and believe differently as they learn to market their place and its, and their, identity.
>
> *(Rothman 1998, 11–12)*

Local tourism workers are valued when they "mirror onto the guest what the visitor wants from them and from their place in a way that affirms the visitor's self-image" (Rothman 1998, 12). Skeptical of the true level of openness to authentic experience of cosmopolitan travelers and their ilk, Rothman notes that the goal of tourism "is not experience but fulfillment—making the chooser feel important, strong, powerful, a member of the right crowd, or whatever else they crave" (Rothman 1998, 14). Rothman believes that the very people who seek out unique, authentic experiences do so to differentiate themselves from others—as smarter or more morally upright. This is not sounding so good for the Indigenous folks whose foodways are to be experienced by tourists.

Some critics, like Grey and Newman, feel that the multicultural appreciation of cosmopolitan consumers is fundamentally incompatible with Indigenous food sovereignty. Such frameworks for gastronomy, they argue, promote local food systems at the expense of Indigenous people, whose cuisines end up either excluded or appropriated. They critique the sleight of hand whereby the "central pillar of [Indigenous] foodways—Indigenous lands" (Grey and Newman 2018, 717) is eliminated while their culinary heritage is alienated and/or commodified—quite apt in the case of US national parks, many of which are built on dispossessed lands. In the face of culinary colonialism, Grey and Newman see both sharing and withholding food as political acts (2018, 717), and they support "the right to hold gastronomic capital back from the market" (2018, 717–18).

Grey and Newman point out that the "cosmopolitan localism" that balances regional food strategies with the consumption of fairly traded goods confuses the positionality of ethnic minorities, oriented toward inclusion and representation, with that of national minorities (Indigenous people), with their claim to specific lands and self-determination; in fact, Grey and Newman argue, they have very different interests (2018, 718). Grey and Newman argue that "Indigenous control over

Indigenous foods is always subordinated" (2018, 719) and critique the settler appropriation of Indigenous gastronomy wherein "Indigenous cuisines are … reoriented toward the demographic that originally sought their eradication" (2018, 719). One might gather that they'd be relieved to find little trace of Indigenous foodways in today's corporate cafeteria spaces. But isn't this overdetermined? Aren't examples like Sean Sherman or Indigikitchen valorizing Indigenous foodways on Indigenous terms? Aren't they connecting cultures, political struggles, and foods, as when Sherman says #86colonialism? Aren't they controlling how their dishes are interpreted, as when Mariah Gladstone writes the script for her Airbnb experiences? Grey and Newman seem suspicious, given that these efforts are beholden to the interests of foodie cosmopolitans, whose "attention tends to wander, creating instability and additional risk for Indigenous communities" (Grey and Newman 2018, 722).

Another reason not to overinvest in culinary tourism as a solution to problems of unsustainability and Indigenous dispossession is structural: Tourism profit margins are low, and volume is necessary for profitability (Bryan 2003, 141). And high-volume tourism requires sameness and standardization, "a reliable product that meets universal standards, despite the dispersal of that product across many widely separated locations" (Kirshenblatt-Gimblett 1998, 152). Some critics have decried the existence of restaurants and food experiences—fancy restaurants, chuck wagon dinners, and daily breakfast rides—in national parks as "lures that have nothing to do with facilitating an experience of the natural resources around which the area has been established" (Sax 1980, 88). They are derided as mere amusements with several shortcomings: "Their capacity to get visitors deep into the park experience seems minimal, they have a mass production quality about them, and they have a considerable capacity to detract attention from the fashioning of a personal agenda" (Sax 1980, 90). But this critique seems to miss how much more the elegant restaurants are aiding with the park's interpretive mission, facilitating an experience of local foods, than "unpretentious restaurants" of the typical fast-food order do, especially when these are run by multinational conglomerates. What if these food experiences were reimagined as the kinds that involve labor and connection to the land and the locals? Might these experiences provide the tourist a distinctive sense of the destination and benefit the local population, goals Sax endorses (1980, 107)? Mantell argues that only concessions that serve an interpretive function should be allowed to expand (1979, 49); to that end, it is worthwhile to consider the possible benefits of integrating commensality into park-located Indigenous food experiences.

Commensality and Shared Food Experiences

One might argue that a cosmopolitan outlook dooms any efforts at true commensality. As much as it may motivate both bioregionalist and Indigenous-centered food experiences in park spaces, cosmopolitanism is not exactly the *sine qua non* of optimal worldviews; for one thing, it tends to get hung up on finding authenticity in ways that risk objectifying Indigenous Others. But there's a way to look at the benefits that cosmopolitanism brings by shifting away from authenticity to *sincerity*. Taylor argues for this move, which

> offers the basis for a shift in *moral* perspective: away from that which would locate touristic value in the successful re-production of 'objective truths'—authenticities—and towards a view of tourism as embodying communicative events involving values important both to the social actors involved, and in themselves.
>
> *(2001, 8–9)*

In a sincerity framework, value is generated from the moment of interaction and negotiation, rather than from an object that is objectified for its assumed authenticity (Taylor 2001, 24). The national parks may be just the spot for this kind of a framework, aligned as they are with nature tourism, which Wang suggests similarly involves a quest for "existential authenticity" rather than the authenticity of objects (1999, 351).

Why does cosmopolitanism desire authenticity? Some suggest that it is because people have become alienated from nature in an age of mass culture and industrialization (Sims 2009, 325), and this allows us to see a direct through-line back to the Romantics who argued for the founding of the national parks. But we usually focus on cosmopolitan obsession with the authenticity of the artifact—for instance, the food, but it might be more productive to focus on "authentification," the process of making claims for authenticity that serve some interests and not others (Jackson 1999).

Cosmopolitanism values travel because the experience cultivates a kind of relational or existential authenticity (Wang 1999, 364). Given that the authenticity-focused discourses of cosmopolitanism are most likely to be taken up by influencers, tastemakers, and cultural intermediaries with the power to make their version of reality count as true, is it possible that these tastes can spur *real* and meaningful social, political, and economic change? Could the life politics of slow travel practices, "informed by a diverse range of ethical sensibilities that bring together pleasurable modes of engaging with nature, or culture, and a politically reflexive sense of identity that is critically aware of the impact of one's own tourist behaviour" (Fullagar, Wilson, and Markwell 2012, 4), spur systemic change that repairs environmental degradation and Indigenous dispossession? Cosmopolitan tastes might, in fact, result in the kinds of appropriate tourism activities described by Bryan in their study of Navajo cultural tourism that "are governed by the uniqueness of the human and natural environments within which the activity takes place" and that enhance the local economy and the quality of the natural and human environment (Bryan 2003, 141).

In his work on the design of sustainable food experiences, Leer suggests that attention should be paid to how the travel-based experience can inspire more sustainable consumption beyond the immediate context of the experience (2020, 65). Part of this is making travelers aware of alternative food systems and challenges to capitalist consumer culture through active participation, disrupting the global inequities between "pleasing local provider and passive tourist consumer" (Leer 2020, 69). He praises slow culinary tourism experiences where travelers take an active part in producing food: "The guest is made responsible and expected to invest more than cash in order to access the full experience" (Leer 2020, 79). He applauds how tourist engagement in physical labor reinforces the idea that everyone needs to take action to achieve sustainability and appreciates the connection between the simple, accessible food and the local context.

> It was also easy to relate to and reproduce at home wherever you might come from in the world. This also highlights how sustainable food experiences might be better fitted to this type of casual café context than to a fine dining context
>
> *(Leer 2020, 80)*

that tends to be centered on the chef, not the collective. Leer's ideas offer some direction for food experiences in the parks.

Heldke, a sharp critic of culinary colonialism, seeks

> a location from which to live food lives that sustain soil and community, that engage with cultural practices other than those we call our own, and that also enable us to live as 'critical eaters,' prepared to make politico-aesthetic decisions through our food choices.
>
> *(Heldke 2006, 4)*

What would this third way, a route between localism (with the appealing localistic thinking, but without the provincialism) and cosmopolitanism (with the productive cross-cultural exchanges, but without the obsessions for authenticity, novelty, and problematic view of the Exotic Other), look like in national park spaces? What food praxes could be there that support both social justice and environmental sustainability?

Heldke is hopeful about the possibilities of culinary cultural exchange to genuinely show a way beyond the problems caused by both cosmopolitanism and localism:

> Cooks who rethink and reshape traditional cuisines in light of the new agricultural contexts in which they find themselves represent one example; rather than insisting on an 'authenticity' that depends upon products made possible only through a vast, energy-expensive global trade network, they make choices that turn their 'exotic' cuisines into something like 'local food.' The 'exotic stranger' who plans to stick around and open a restaurant can and should think about the ways in which their cuisine can meet the expectations of the land in which it is now situated.
>
> *(Heldke 2006, 21)*

Commensal experiences of culinary cultural tourism in the parks satisfy cosmopolitan travelers. However, their true promise lies elsewhere: Beyond simply making for a good vacation, they are part of a profound reimagining of social, political, and economic relations among people and to the land.

Research Directions: The Future of Eating in National Parks

In conclusion, it is fair to problematize aspects of cosmopolitan taste and desire, and to proceed cautiously regarding food enterprises that aim to satisfy cosmopolitan travelers hungry for authenticity and a certain type of self-making. I have attempted in this book to examine closely the affordances and constraints of catering to cosmopolitanism in park eating as a vehicle for environmental and social justice. I conclude that national parks eating experiences that center Indigenous people and their foodways, and that more broadly subscribe to tenets of bioregionalism, are productive areas for future research and investment.

As this research develops, observational ethnographic data and description of the variety of ways of eating in the park will be important. Camp cooking and picnicking are worthy areas for further investigation, as are subsistence hunting and gathering. At present, the everyday experiences of these ways of eating in the parks are not well documented,[2] and thus, they don't lend themselves very well to the kind of textual analysis primarily employed in this book. In addition, the growing scholarly literature offering comparative analyses of national park governance in international contexts could be expanded to consider more deeply what commitments to the land and to the people are enshrined in the food experiences these parks offer. Finally, as we are seeing a shift toward catering to cosmopolitan desires in the eating experiences available in the parks, research both conceptual and empirical might consider the implications for non-cosmopolitan eaters. Does a rising tide of concern for environmental and social justice that finds its expression in park eating experiences indeed lift all boats, or might expense and access issues further the marginalization of those who already find the parks inaccessible or not compelling? Research that considers the implications for non-cosmopolitans of this shift toward catering to cosmopolitans in park eating is important and welcome.

Notes

1 Little information is publicly available about national park restaurant art curation practices. However, a website by artist Natasha Bacca explains the process of having her work exhibited in the Metate Room; she mentions being told by the site manager that her chromogenic photogram of a wine glass would fit in well with the theme of "southwest design." Southwest becomes a proxy for Indigenous. See http://natashabacca.weebly. com/blog/artwork-by-natasha-bacca-at-mesa-verde-national-park.

2 There are a few cookbooks in print featuring outdoor cooking recipes from the national parks, and a smattering of instructional YouTube videos about camp stove cooking in the parks, but I am not aware of a more sustained set of texts pertaining to national park cooking.

References

AIANTA. 2023. "American Indians and Route 66." Brochure Produced by the American Indian Alaska Native Tourism Association. Accessed July 18, 2023. https://www.aianta.org/wp-content/uploads/2020/03/American_Indians_Route66.pdf.

Airbnb. 2022. "Dinner Featuring Local Blackfeet Foods." Accessed November 11, 2022. https://abnb.me/xKHDiDlxltb.

Andruss, Van, Christopher Plant, Judith Plant, and Eleanor Wright, ed. 1990. *Home! A Bioregional Reader.* Gabriola: New Society Publishers.

Barsh, Russel L. 1999. "Chronic Health Effects of Dispossession and Dietary Change: Lessons from North American Hunter-Gatherers." *Medical Anthropology* 18 (2): 135–61.

Beard-Moose, Christina Taylor. 2011. *Public Indians, Private Cherokees: Tourism and Tradition on Tribal Ground.* Tuscaloosa: University of Alabama Press.

Bigley, James II. 2016. "Meet the People Behind These Nine Cuyahoga Valley National Park Farms." *Cleveland Magazine*, August 2. https://clevelandmagazine.com/food-drink/articles/meet-the-people-behind-these-nine-cuyahoga-valley-national-park-farms.

Bourdieu, Pierre. 1984. *Distinction: A Social Critique of the Judgement of Taste.* Translated by Richard Nice. New York: Routledge and Kegan Paul.

Bryan, William L., Jr. 2003. "Appropriate Cultural Tourism—Can It Exist? Searching for an Answer: Three Arizona Case Studies." In *The Culture of Tourism, the Tourism of Culture: Selling the Past to the Present in the American Southwest*, edited by Hal K. Rothman, 140–63. Albuquerque: University of New Mexico Press.

Burnham, Philip. 2013. *Indian Country, God's Country: Native Americans and the National Parks.* Bloomington, IN: Authors Guild/iUniverse.

Campbell, Colin. 1987. *The Romantic Ethic and the Spirit of Modern Consumerism.* Oxford: Blackwell.

Campbell, Colin. 2005. "The Craft Consumer: Culture, Craft and Consumption in a Postmodern Society." *Journal of Consumer Culture* 5 (1): 23–42.

Carr, Anna, Lisa Ruhanen, and Michelle Whitford. 2016. "Indigenous Peoples and Tourism: The Challenges and Opportunities for Sustainable Tourism." *Journal of Sustainable Tourism* 24 (8–9): 1067–79. https://doi.org/10.1080/09669582.2016.1206112.

Carr, Ethan. 2007. *Mission 66: Modernism and the National Park Dilemma.* Amherst: University of Massachusetts Press.

Catton, Theodore. 1997. *Inhabited Wilderness: Indians, Eskimos, and National Parks in Alaska.* Albuquerque: University of New Mexico Press.

Clancy, Michael, ed. 2017. *Slow Tourism, Food, and Cities: Pace and the Search for the 'Good Life.'* New York: Routledge.

Concession Contracts. 1998. "Public Law 105-391 of November 13, 1998, National Parks Omnibus Management Act of 1998." Code of Federal Regulations, part 51 (2022): 329–59. https://www.govinfo.gov/content/pkg/CFR-2021-title36-vol1/pdf/CFR-2021-title36-vol1-part51.pdf.

Conn, Jennifer. 2022. "Countryside's Howe Meadow Farmers' Markets Return to Cuyahoga Valley National Park Saturday." *SpectrumNews1*, May 6. https://spectrumnews1.com/oh/columbus/community/2022/05/05/countryside-howe-meadow-farmers—markets-cuyahoga-valley-national-park.

Cothran, Boyd. 2010. "Working the Indian Field Days: The Economy of Authenticity and the Question of Agency in Yosemite Valley." *American Indian Quarterly* 34 (2): 194–223.

Countryside. 2022. "The Countryside Initiative." *Countrysidefoodandfarms.org.* Accessed September 8, 2022. https://countrysidefoodandfarms.org/countryside-initiative/.

de la Barre, Suzanne and Patrick Brouder. 2013. "Consuming Stories: Placing Food in the Arctic Tourism Experience." *Journal of Heritage Tourism* 8 (2–3): 213–23.

Delaware North. 2022. "Culinary Events at Shenandoah National Park." *GoShenandoah. com*. Accessed September 8, 2022. https://www.goshenandoah.com/activities-events/ culinary-events.

de Solier, Isabelle. 2013. *Food and the Self: Consumption, Production and Material Culture*. London: Bloomsbury.

Diamond, Jared. 1992. "Diabetes Running Wild." *Nature* 357: 362–3. https://doi.org/10. 1038/357362a0.

Eastern National. 2023. "Homepage." Accessed July 18, 2023. https://easternnational.org/.

Finn, S. Margot. 2017. *Discriminating Taste: How Class Anxiety Created the American Food Revolution*. New Brunswick: Rutgers University Press.

Fullagar, Simone, Erica Wilson, and Kevin Markwell. 2012. "Starting Slow: Thinking Through Slow Mobilities and Experiences." In *Slow Tourism: Experiences and Mobilities*, edited by Simone Fullagar, Kevin Markwell, and Erica Wilson, 1–8. Bristol: Channel View.

Fusté-Forné, Francesc and Tazim Jamal. 2020. "Slow Food Tourism: An Ethical Microtrend for the Anthropocene." *Journal of Tourism Futures* 6 (3): 227–32. https://doi. org/10.1108/JTF-10-2019-0120.

Gladstone, Mariah. Personal communication (meeting). Old Faithful, Yellowstone National Park, September 15, 2022.

Glanz, Karen, Michael Basil, Edward Maibach, Jeanne Goldberg, and Dan Snyder. 1998. "Why Americans Eat What They Do." *Journal of the American Dietetic Association* 98 (10): 1118–26.

Golden Gate National Parks Conservancy. 2022. "Food in the Parks." *ParksConservancy. org*. Accessed September 8, 2022.

Grey, Sam and Lenore Newman. 2018. "Beyond Culinary Colonialism: Indigenous Food Sovereignty, Liberal Multiculturalism, and the Control of Gastronomic Capital." *Agriculture and Human Values* 35: 717–30. https://doi.org/10.1007/s10460-018-9868-2.

Grey, Sam and Rauna Kuokkanen. 2020. "Indigenous Governance of Cultural Heritage: Searching for Alternatives to Co-Management." *International Journal of Heritage Studies* 26 (10): 919–41. https://doi.org/10.1080/13527258.2019.1703202.

Guyette, Susan and David White. 2003. "Reducing the Impacts of Tourism Through Cross-Cultural Planning." In *The Culture of Tourism, the Tourism of Culture: Selling the Past to the Present in the American Southwest*, edited by Hal K. Rothman, 164–84. Albuquerque: University of New Mexico Press.

Hall, C. Michael and Liz Sharples. 2003. "The Consumption of Experiences or the Experience of Consumption? An Introduction to the Tourism of Taste." In *Food Tourism Around the World: Development, Management and Markets*, edited by C. Michael Hall, Liz Sharples, Richard Mitchell, Niki Macionis, and Brock Cambourne, 1–24. Burlington, MA: Butterworth Heinemann.

Hashimoto, Atsuko and David J. Telfer. 2015. "Culinary Trails." In *Heritage Cuisines: Traditions, Identities, and Tourism*, edited by Timothy J. Dallen, 132–47. London: Routledge.

Heldke, Lisa. 2006. "Beyond Cosmopolitanism and Localism." *Appetite* 47 (3): 390. https:// doi.org/10.1016/j.appet.2006.08.025.

Heldke, Lisa. 2012. "Down-Home Global Cooking: A Third Option between Cosmopolitanism and Localism." In *The Philosophy of Food*, edited by David M. Kaplan, 33–51. Berkeley: University of California Press. https://books.google.com/books?hl=en&lr=& id=idubdeClka0C&oi=fnd&pg=PA33&dq=bioregionalism+culinary+tourism&ots=r1p XVj_iuG&sig=WC7CUaSxvS_9mtvYaJobmy4qhAY#v=onepage&q&f=false.

Hinch, Tom and Richard R. Butler. 1996. "Indigenous Tourism: A Common Ground for Discussion." In *Tourism and Indigenous Peoples*, edited by Richard E. Butler and Tom Hinch, 3–22. London: Thomson International Business Press.

Humboldt County Visitors Bureau. 2022a. "Arcata Farmers Market." Accessed September 8, 2022. http://www.visitredwoods.com/event/arcata-farmers-market/3348/.

Humboldt County Visitors Bureau. 2022b. "Redwood Coast 3 Day Culinary Tour." Accessed September 8, 2022. https://www.visitredwoods.com/listing/redwood-coast-3-day-culinary-tour/182/.

IndigiKitchen. n.d. "Native Health." Accessed November 11, 2022. https://www.indigikitchen.com/.

Institute at the Golden Gate. 2011. "Food for the Parks: Case Studies of Sustainable Food in America's Most Treasured Places." Accessed September 9, 2022. https://www.parksconservancy.org/conservation/food-parks.

Institute at the Golden Gate. 2012. "Food for the Parks: A Roadmap to Success." Accessed September 9, 2022. https://www.parksconservancy.org/conservation/food-parks.

Intertribal Agriculture Council. n.d. "American Indian Foods." Accessed November 12, 2022. https://www.indianagfoods.org/american-indian-trademark.

Jackson, Peter. 1999. "Commodity Cultures: The Traffic in Things." *Transactions of the Institute of British Geographers* 24 (1): 95–108.

Joe, Jennie R. and Robert S. Young, ed. 1994. *Diabetes as a Disease of Civilization: The Impact of Culture Change on Indigenous Peoples*. New York: Mouton de Gruyter.

Keller, Robert H. and Michael F. Turek. 1998. *American Indians and National Parks*. Tucson: University of Arizona Press.

Kim, Yeong Gug, Anita Eves, and Caroline Scarles. 2009. "Building a Model of Local Food Consumption on Trips and Holidays: A Grounded Theory Approach." *International Journal of Hospitality Management* 28 (3): 423–31.

Kirshenblatt-Gimblett, Barbara. 1998. *Destination Culture: Tourism, Museums, and Heritage*. Berkeley: University of California Press.

Kirshenblatt-Gimblett, Barbara. 2004. "Foreword." In *Culinary Tourism*, edited by Lucy M. Long, xi–xiv. Lexington: University Press of Kentucky.

Knorr, Beth. 2021. "Sustainable Farming." Audio file. Countryside Food and Farms, Cuyahoga Valley National Park. Posted September 22, 2021. https://www.nps.gov/cuva/learn/historyculture/countryside.htm.

Kormann, Carolyn. 2022. "How Owamni Became the Best New Restaurant in the United States." *The New Yorker*, September 12. https://www.newyorker.com/magazine/2022/09/19/how-owamni-became-the-best-new-restaurant-in-the-united-states.

Leer, Jonatan. 2020. "Designing Sustainable Food Experiences: Rethinking Sustainable Food Tourism." *International Journal of Food Design* 5 (1–2): 65–82.

Lewis, Courtney. 2019. *Sovereign Entrepreneurs: Cherokee Small-Business Owners and the Making of Economic Sovereignty*. Chapel Hill: University of North Carolina Press.

Long, Lucy M. 2004. "Introduction." In *Culinary Tourism*, edited by Lucy M. Long, 1–19. Lexington: University Press of Kentucky.

MacCannell, Dean. 1973. "Staged Authenticity: Arrangements of Social Space in Tourist Settings." *American Journal of Sociology* 79 (3): 589–603.

MacCannell, Dean. 2011. *The Ethics of Sightseeing*. Berkeley: University of California Press.

Mantell, Michael. 1979. "Preservations and Use: Concessions in the National Parks." *Ecology Law Quarterly* 8 (1): 1–54.

Marsh, Jill and Gwynne Trivelpiece. 2022. "Xanterra's Native American Buying Practices 2022." Presented at the 3rd Native American Handicrafts Discussion Forum, May 25, 2022. https://www.nps.gov/subjects/concessions/upload/3rd-NAH-Discussion-Forum-2022-05-25_Xanterra.pdf.

Mason, Robb and Barry O'Mahony. 2011. "On the Trail of Food and Wine: The Tourist Search for Meaningful Experience." *Annals of Leisure Research* 10 (3–4): 498–517.

Meethan, Kevin. 2015. "Making the Difference: The Experience Economy and the Future of Regional Food Tourism." In *The Future of Food Tourism: Foodies, Experiences, Exclusivity, Visions and Political Capital*, edited by Ian Yeoman, Una McMahon-Beattie, Kevin Fields, Julia N. Albrecht, and Kevin Meethan, 114–26. Bristol: Channel View.

Mesa Verde Association. n.d. "Who We Are." Accessed November 8, 2022. https://www.mesaverde.org/who-we-are.

Mihesuah, Devon A. 2019. "Nephi Craig: Life in Second Sight." In *Indigenous Food Sovereignty in the United States: Restoring Cultural Knowledge, Protecting Environments, and Regaining Health*, edited by Devon A. Mihesuah and Elizabeth Hoover, 300–19. Norman: University of Oklahoma Press.

Morris, Haydon and Michael Romeril. 1986. "Farm Tourism in England's Peak National Park." *Environmentalist* 6: 105–10. https://doi.org/10.1007/BF02277233.

Munt, Ian. 1994. "The 'Other' Postmodern Tourism: Culture, Travel and the New Middle Classes." *Theory, Culture and Society* 11 (3): 101–23.

Nalewicki, Jennifer. 2021. "The Fruits of Fruita: Preserving a Legacy in Utah's Capitol Reef National Park." *Borderlore.org*, October 28. https://borderlore.org/the-fruits-of-fruita/.

NATIFS. n.d. "North American Traditional Indigenous Food Systems." Accessed November 11, 2022. https://www.natifs.org/.

National Park Foundation. 2016. "4 National Parks for the Foodie in You." *FindYourPark.com*. Accessed September 8, 2022. https://findyourpark.com/get-inspired/4-national-parks-foodie-you.

National Park Service. 2015. "Tourism to Olympic National Park Creates $365,559,900 in Economic Benefits." Olympic National Park, April 23. Accessed September 8, 2022. https://www.nps.gov/olym/learn/news/tourism-to-olympic-national-park-creates-millions-economic-benefits.htm.

National Park Service. 2016a. "Economic Impacts of National Park Visitation Increase." Redwood National Park, July 15. https://www.nps.gov/redw/learn/news/econ.htm.

National Park Service. 2016b. "Tribal Tourism and Native Voices in Parks." Last updated July 16, 2022. https://www.nps.gov/articles/2016npstribaltourismhighlights.htm.

National Park Service. 2020. "Places." Martin Van Buren National Historic Site, September 25, 2020. https://www.nps.gov/mava/learn/historyculture/places.htm.

National Park Service. 2022a. "Procuring Local Foods." Posted May 10, 2022. https://www.nps.gov/articles/000/local-foods.htm.

National Park Service. 2022b. "Authentic Native Handicrafts." Posted June 28, 2022. https://www.nps.gov/subjects/concessions/anh.htm.

National Park Service. 2022c. "National Park Service Issues New Policy Guidance to Strengthen Tribal Co-Stewardship of National Park Lands and Waters." Posted September 13, 2022. https://www.nps.gov/orgs/1207/national-park-service-issues-new-policy-guidance-to-strengthen-tribal-co-stewardship-of-national-park-lands-and-waters.htm.

National Park Service. 2022d. "Countryside Food and Farms." Cuyahoga Valley National Park, November 1, 2021. https://www.nps.gov/cuva/learn/historyculture/countryside.htm.

Nelson, Kate. 2021. "The Man Who Sees a Future Where Indigenous Foods Are as Ubiquitous as Burgers." *Esquire*, May 20. https://www.esquire.com/food-drink/restaurants/a36474711/chef-sean-sherman-owamni-indigenous-minneapolis-restaurant-profile/.

Nielsen, Noah and Erica Wilson. 2012. "From Invisible to Indigenous-Driven: A Critical Typology of Research in Indigenous Tourism." *Journal of Hospitality and Tourism Management* 19 (1): 67–75. https://doi.org/10.1017/jht.2012.6.

Parasecoli, Fabio and Paulo de Abreu e Lima. 2012. "Eat Your Way Through Culture: Gastronomic Tourism as Performance and Bodily Experience." In *Slow Tourism: Experiences and Mobilities*, edited by Simone Fullagar, Kevin Markwell, and Erica Wilson, 69–83. Bristol: Channel View.

Park Ranger John. 2022. "Capitol Reef National Park: Epic Guide 2022." Accessed September 8, 2022. https://www.parkrangerjohn.com/capitol-reef-national-park/.

Pfeiffer, Catherine, Roel A. Jongeneel, Marthijn P.W. Sonneveld, and Jetse J. Stoorvogel. 2009. "Landscape Properties as Drivers for Farm Diversification: A Dutch Case Study." *Land Use Policy* 26 (4): 1106–15.

Pine II, B. Joseph and James H. Gilmore. 1999. *The Experience Economy: Work is Theatre and Every Business a Stage*. Boston, MA: Harvard Business School Press.

Pratt, Mary Louise. 1992. *Imperial Eyes: Travel Writing and Transculturation*. New York: Routledge.

Prentice, Richard. 2001. "Experiential and Cultural Tourism: Museums and the Marketing of the New Romanticism of Evoked Authenticity." *Museum Management and Curatorship* 19 (1): 5–26.

Repanshek, Kurt. 2009. "Dining at the Parks: Mesa Verde National Park's Chef Ensures the Southwest Flows through His Dishes." *National Parks Traveler.org*, July 18. https://www.nationalparkstraveler.org/2009/07/dining-parks-mesa-verde-national-parks-chef-ensures-southwest-flows-through-his-dishes.

Rogers, Jerry L. 2009. "Ethnography in a National Park Service Second Century." The National Park Service Centennial Essay Series: 1916–2016. *The George Wright Forum* 26 (3). http://www.georgewright.org/263rogers.pdf.

Rothman, Hal K. 1998. *Devil's Bargains: Tourism in the 20th Century American West*. Lawrence: University of Kansas Press.

Roxbury Farm CSA. 2022. "Roxbury Farm and Secure Land Access." Accessed September 8, 2022. https://www.roxburyfarm.com/roxbury-farm-and-secure-land-tenure.

Sax, Joseph L. 1980. *Mountains Without Handrails: Reflections on the National Parks*. Ann Arbor: University of Michigan Press.

Scarpato, Rosario and Roberto Daniele. 2003. "New Global Cuisine: Tourism, Authenticity and Sense of Place in Postmodern Gastronomy." In *Food Tourism Around the World: Development, Management and Markets*, edited by C. Michael Hall, Liz Sharples, Richard Mitchell, Niki Macionis, and Brock Cambourne, 296–313. Burlington, MA: Butterworth Heinemann.

Scherl, Lea M. 2005. "Protected Areas and Local and Indigenous Communities." In *Friends for Life: New Partnerships in Support of Protected Areas*, edited by Jeffrey A. McNeely, 102–12. Gland: International Union for Conservation of Nature.

Scherl, Lea M. and Stephen Edwards. 2007. "Tourism, Indigenous and Local Communities and Protected Areas in Developing Nations." In *Tourism and Protected Areas: Benefits Beyond Boundaries*, edited by Robyn Bushell and Paul F.J. Eagles, 71–88. Wallingford: CAB International.

Senate Hearing 115–75. 2017. "Cultural Sovereignty Series: Modernizing the Indian Arts and Crafts Act to Honor Native Identity and Expression." Field hearing before the Committee on Indian Affairs, US Senate, July 7. https://www.govinfo.gov/content/pkg/CHRG-115shrg26821/html/CHRG-115shrg26821.htm.

Sharples, Liz. 2003. "Food Tourism in the Peak District National Park, England." In *Food Tourism Around the World: Development, Management and Markets*, edited by C. Michael Hall, Liz Sharples, Richard Mitchell, Niki Macionis, and Brock Cambourne, 206–27. Burlington, MA: Butterworth Heinemann.

Sherman, Sean. 2019. "Voices from the Indigenous Food Movement: Sean Sherman." In *Indigenous Food Sovereignty in the United States: Restoring Cultural Knowledge, Protecting Environments, and Regaining Health*, edited by Devon A. Mihesuah and Elizabeth Hoover, 44–48. Norman: University of Oklahoma Press.

Sims, Rebecca. 2009. "Food, Place and Authenticity: Local Food and the Sustainable Tourism Experience." *Journal of Sustainable Tourism* 17 (3): 321–36. https://doi.org/10.1080/09669580802359293.

Sioux Chef. n.d. "We are The Sioux Chef." Accessed November 11, 2022. https://sioux-chef.com/.

Sleeping Rainbow Adventures. 2022. "Best Things to Do in Capitol Reef." Accessed September 8, 2022. https://sradventures.com/about-us/best-things-to-do-in-capitol-reef/.

Slocum, Susan L. and Kynda R. Curtis. 2016. "Assessing Sustainable Food Behaviours of National Park Visitors: Domestic/On Vocation Linkages, and Their Implications for Park Policies." *Journal of Sustainable Tourism* 24 (1): 153–67. https://doi.org/10.1080/09669582.2015.1062019.

Spence, Mark. 1999. *Dispossessing the Wilderness: Indian Removal and the Making of the National Parks*. New York: Oxford University Press.

Staiff, Russell, Robyn Bushell, and Peter Kennedy. 2002. "Interpretation in National Parks: Some Critical Questions." *Journal of Sustainable Tourism* 10 (2): 97–113. https://doi.org/10.1080/09669580208667156.

Stevens, Alison. 2020. "Taste of Place and Provenance." WWU Honors College Senior Projects. 349. https://cedar.wwu.edu/wwu_honors/349.

Taylor, John P. 2001. "Authenticity and Sincerity in Tourism." *Annals of Tourism Research* 28 (1): 7–26.

Thomsen, Jennifer M., Robert B. Powell, and Diana Allen. 2013. "Park Health Resources: Benefits, Values, and Implications." *Park Science* 30 (2): 30–36. https://www.researchgate.net/profile/Robert-Powell-4/publication/263806493_What_is_the_value_and_implications_of_viewing_park_resources_as_health_resources/links/02e7e53c6b2f9eefce000000/What-is-the-value-and-implications-of-viewing-park-resources-as-health-resources.pdf.

Treuer, David. 2021. "Return the National Parks to the Tribes." *The Atlantic* 327 (4): 30–45.

Tuck, Eve and K. Wayne Yang. 2012. "Decolonization Is Not a Metaphor." *Decolonization: Indigeneity, Education and Society* 1 (1): 1–40.

Visit Durango. 2022. "Mesa Verde National Park Facts." Accessed August 26, 2022. https://www.durango.org/press-room/fact-sheets/mesa-verde-national-park-facts/.

Wang, Ning. 1999. "Rethinking Authenticity in Tourism Experience." *Annals of Tourism Research* 26 (2): 349–70.

Watson, Rebecca. 2022. "Parks That Celebrate Native American Heritage." National Park Foundation. Accessed November 7, 2022. https://www.nationalparks.org/connect/blog/parks-celebrate-native-american-heritage.

Wilkinson, Todd. 1996. "Native Americans Challenge Park Agency for Land Rights." *The Christian Science Monitor*, October 22. https://www.csmonitor.com/1996/1022/102296.us.us.3.html.

Xanterra. 2022. "Green Restaurant Certifications." Accessed November 9, 2022. https://www.xanterra.com/stories/green-restaurant-certifications/.

Yazzie, Brian. 2019. "Voices from the Indigenous Food Movement: Brian Yazzie." In *Indigenous Food Sovereignty in the United States: Restoring Cultural Knowledge, Protecting Environments, and Regaining Health*, edited by Devon A. Mihesuah and Elizabeth Hoover, 53–56. Norman: University of Oklahoma Press.

Yeoman, Ian and Una McMahon-Beattie. 2015. "The Future of Food Tourism: The Star Trek Replicator and Exclusivity." In *The Future of Food Tourism: Foodies, Experiences, Exclusivity, Visions and Political Capital*, edited by Ian Yeoman, Una McMahon-Beattie, Kevin Fields, Julia N. Albrecht, and Kevin Meethan, 23–45. Bristol: Channel View.

Yeoman, Ian, Una McMahon-Beattie, Kevin Fields, Julia N. Albrecht, and Kevin Meethan, eds. 2015. *The Future of Food Tourism: Foodies, Experiences, Exclusivity, Visions and Political Capital*. Bristol: Channel View.

Zeppel, Heather. 2009. "National Parks as Cultural Landscapes: Indigenous Peoples, Conservation, and Tourism." In *Tourism and National Parks: Perspectives on Development, Histories, and Change*, edited by Warwick Frost and C. Michael Hall, 259–81. New York: Routledge.

INDEX

Note: **Bold** page numbers refer to tables and page numbers followed by "n" denote endnotes.